I0056194

Titanate and Titania Nanotubes
Synthesis, Properties and Applications

RSC Nanoscience & Nanotechnology

Series Editors:
Professor Paul O'Brien, *University of Manchester, UK*
Professor Sir Harry Kroto FRS, *University of Sussex, UK*
Professor Harold Craighead, *Cornell University, USA*

How to obtain future titles on publication:
A standing order plan is available for this series. A standing order will bring delivery of each new volume immediately on publication.

For further information please contact:
Book Sales Department, Royal Society of Chemistry,
Thomas Graham House, Science Park, Milton Road, Cambridge,
CB4 0WF, UK
Telephone: +44 (0)1223 420066, Fax: +44 (0)1223 420247, Email: books@rsc.org
Visit our website at http://www.rsc.org/Shop/Books/

Titanate and Titania Nanotubes
Synthesis, Properties and Applications

Dmitry V. Bavykin and Frank C. Walsh
School of Engineering Sciences, University of Southampton, Southampton, UK

RSCPublishing

RSC Nanoscience & Nanotechnology No. 12

ISBN: 978-1-84755-910-4
ISSN: 1757-7136

A catalogue record for this book is available from the British Library

© Dmitry V. Bavykin and Frank C. Walsh 2010

All rights reserved

Apart from fair dealing for the purposes of research for non-commercial purposes or for private study, criticism or review, as permitted under the Copyright, Designs and Patents Act 1988 and the Copyright and Related Rights Regulations 2003, this publication may not be reproduced, stored or transmitted, in any form or by any means, without the prior permission in writing of The Royal Society of Chemistry or the copyright owner, or in the case of reproduction in accordance with the terms of licences issued by the Copyright Licensing Agency in the UK, or in accordance with the terms of the licences issued by the appropriate Reproduction Rights Organization outside the UK. Enquiries concerning reproduction outside the terms stated here should be sent to The Royal Society of Chemistry at the address printed on this page.

The RSC is not responsible for individual opinions expressed in this work.

Published by The Royal Society of Chemistry,
Thomas Graham House, Science Park, Milton Road,
Cambridge CB4 0WF, UK

Registered Charity Number 207890

For further information see our web site at www.rsc.org

Dedications

From Dmitry Bavykin
   to my first research supervisor: Professor E. N. Savinov
   (15 January 1954–13 November 2002)

From Frank Walsh
   to Pamela, Steven and Heather, my understanding wife, son and daughter.

Preface

"Good things come in small packages"
A proverbial saying

"Less is more"
Robert Browning in a 1855 poem; the phrase was then popularized
by the architect Ludwig Mies van der Rohe (27 March 1886–17 August
1969)

The Importance of Nanostructured Titanates

Nanostructured materials have been with us for many centuries. In 1991, a
paper by Sumio Iijima on carbon nanotubes (*Nature*, 1991, **354**, 56–8) stimulated recognition of the importance and the structural elegance of these
materials. This has catalysed an explosion of interest in this field, with thou-
sands of scientific papers on nanomaterials being published over the last
decade.

The synthesis of nanostructured titanium oxide and titanates is often
attributed to Kasuga and co-workers in 1998 (T. Kasuga, M. Hiramatsu, A.
Hoson, T. Sekino, K. Niihara, *Langmuir*, 1998, **14**, 3160–3163). Since 1998,
hundreds of papers and five reviews have appeared on the synthesis and
characterisation of nanostructured titanates and titanium dioxide. For example, the year of 2009 saw over 800 papers on nanotubular titanates and titanium
dioxide. While titanium oxides are far less popular than carbon nanostructures,
they have the marked advantages of low cost and facile synthesis routes which
use conventional laboratory (and scaleable technology) methods.

Elongated structures are of particular importance, since long, thin nanotubes
and nanofibres can provide a high specific surface area in a structured fashion.
In the case of elongated titanium oxide and titanate nanostructures, it is possible to tailor the synthesis conditions to obtain tubes or fibres. A wide range of

RSC Nanoscience & Nanotechnology No. 12
Titanate and Titania Nanotubes: Synthesis, Properties and Applications
By Dmitry V. Bavykin and Frank C. Walsh
© Dmitry V. Bavykin and Frank C. Walsh 2010
Published by the Royal Society of Chemistry, www.rsc.org

synthesis techniques can de deployed, the major ones being hydrothermal treatment, sol–gel processes and the anodising of titanium metal. The attractive properties of titanium oxide nanostructures include: the ease of synthesis; a relatively low cost (especially in comparison to that of carbon nanotubes); a high specific surface area; the ability to functionalise the surface *via* chemisorption and the ion-exchange of species; and the possibility of reversible gas (*e.g.*, hydrogen) adsorption.

Current application areas of titanium oxide nanostructures include: dye-sensitised solar cells; the photocatalytic degradation of environmental contaminants; and solid-state lithium batteries. Other important potential applications are found in the fields of gas- and liquid-phase catalysis, biosensors and the electroplating of composite coatings.

The authors have several aims in writing this book: (*a*) to provide a timely statement concerning the progress made in this field, (*b*) to consolidate knowledge which is currently scattered throughout many scientific journals, and (*c*) to stimulate further research into titanium (and other metal) oxide nanostructures, in order to improve the industrial development of devices and systems which capitalise on their special properties.

This book is divided into five chapters. Chapter 1 sets the scene by concisely reviewing the history of nanomaterials and the emergence of titanium oxide nanostructures, resulting from the immense interest in nanomaterials. A brief review of the wet chemical syntheses of other elongated inorganic nanomaterials is also included. Chapter 2 concerns the methods available for the synthesis of elongated TiO$_2$-based nanostructures, including: the hydrothermal, sol–gel and anodising approaches. The mechanism of nanostructure growth is also treated. Chapter 3 considers the structural and physical properties of elongated titanium oxide nanostructures, as studied by crystallographic, adsorption and electron microscopy imaging. Chapter 4 reviews the chemical properties of titanate nanostructures, the possibility of transforming one nanostructure to another by simple chemical treatments and examples of successful coating techniques. The final chapter considers the actual and potential applications of titanium oxide and titanate nanostructures, with an emphasis on energy conversion, catalysis (chemical, biochemical and electrochemical) and speciality (*e.g.*, electronic and magnetic) materials. The authors have sought to make each chapter self-contained, with its own set of references to support and extend the text.

Dmitry V. Bavykin and Frank C. Walsh
University of Southampton, UK

Contents

RSC Nanoscience & Nanotechnology No. 12
Titanate and Titania Nanotubes: Synthesis, Properties and Applications
By Dmitry V. Bavykin and Frank C. Walsh
© Dmitry V. Bavykin and Frank C. Walsh 2010
Published by the Royal Society of Chemistry, www.rsc.org

About the Authors

The authors have collaborated closely on research into elongated titanate nanostructures since a Royal Society Fellowship award to Dmitry Bavykin to work with Alexei Lapkin and Frank Walsh in the Department of Chemical Engineering, University of Bath in 2002.

Both authors are members of the Electrochemical Engineering Laboratory, the Energy Technology Research Group and the Engineering Materials Research Group at the University of Southampton. Frank Walsh is a member of the national Centre for Advanced Tribology at Southampton and directs the Research Institute for Industry.

Dmitry Bavykin is a physical chemist with interests in the synthesis and physico-chemical characterisation of metal oxide nanomaterials. His research interests are centred on inorganic materials with a focus on metal oxide (especially titanate) nanomaterials. The majority of his research output has concerned elongated titanates (*via* the hydrothermal method) together with their physico-chemical characterisation for applications including catalysis.

He received his MSc degree in Chemistry in Novosibirsk State University (Russian Federation) in 1995, followed by his post-graduate studies in Boreskov's Institute of Catalysis (Novosibirsk, Russian Federation) in the area of CdS nanoparticles under the supervision of Professor Evgeni N. Savinov in the Laboratory of Photocatalysis on Semiconductors. After obtaining his PhD degree in 1998, he continued research in the same laboratory on synthesis, characterisation and photocatalytic studies of titanium dioxide-based materials. Since 2002, his active area of research has become the study of titanate nanotubes, which has resulted in over 25 publications, including two review papers and a paper which was awarded the Johnson Matthey Silver Medal of the Institute of Metal Finishing.

Frank Walsh is an electrochemical engineer and physical chemist with an interest in the physico-chemical and electrochemical characterisation of nanomaterials for applications such as fuel cells, batteries, solar cells and coatings for tribological protection. His diverse research interests include: correlations among reaction conditions, the structure of nanomaterials, their chemical properties and applications.

He was educated at Bootle Grammar School for Boys. A BSc degree in Applied Chemistry (Portsmouth) led to study periods at the Universities of Manchester and Loughborough, resulting in an MSc in Materials Protection (Loughborough) in 1976. Continued research work at Loughborough

University and in industry, supervised by Professor David Gabe (a metallurgist and electrochemical coatings specialist) resulted in a PhD in 1981. These studies were instrumental in a shift towards chemical engineering, and a two-year period in industry was spent designing and evaluating electrode materials, reactors and process plants for electrochemical synthesis and environmental treatment.

In 1983, Frank pursued industrial contract research and consultancy in electrochemical engineering in the Chemistry Department at the University of Southampton. 1986 saw a move to a lectureship in chemical technology at the Department of Pure and Applied Chemistry at the University of Strathclyde, Glasgow, with research activities in electrochemical synthesis, electrode kinetics and synchrotron X-ray analysis of electrode materials. A senior lectureship at the University of Portsmouth in 1988 was followed by promotions to a Readership (1990), a Chair (1992), Head of the Schools of Chemistry, Physics and Radiography (1997) then Pharmacy and Biomedical Sciences (1999). In 2001, a move to a Chair in Electrochemical Engineering at the Department of Chemical Engineering at the University of Bath was followed by the Head of Department position.

In 2004, he returned to the University of Southampton (in the School of Engineering Sciences), which enabled (*a*) the formation of the Electrochemical Engineering Laboratory, (*b*) an expansion of the Research Institute for Industry, (*c*) collaborations with the Electrochemistry and Surface Science Group in the School of Chemistry, (*d*) the formation of an Energy Technology Research Group, and (*e*) contributions to a national Centre for Advanced Tribology at Southampton.

Having crossed several disciplines, Frank is a Chartered- Chemist, Engineer, Environmentalist and Scientist and a Fellow of the Royal Society of Chemistry, the Institute of Materials, Mining and Metallurgy, the Institute of Corrosion and The Institute of Metal Finishing. His research output, which spans the areas of energy conversion, electroactive nanomaterials, coating technology, electrochemical monitoring and sensors, corrosion, surface finishing and electrochemical process engineering, includes: four text books, 70 short course papers, > 200 conference presentations, > 260 research papers and > 50 educational papers. He has written books on Industrial Electrochemistry (with Derek Pletcher, 1990), A First Course in Electrochemical Engineering (1993) and Electrochemistry for Metal Finishers (a distance-learning course, 1995). Frank has developed or improved over 50 industrial processes and trained 60 PhD students.

Frank Walsh is pleased to have been a Visiting Professor at the University of Wollongong in Australia, and currently holds a Visiting Chair in Electrochemical Technology at the University of Strathclyde, Scotland. He was awarded the Westinghouse Prize (1999, 2009) and Johnson Matthey Silver Medal (2007) of the Institute of Metal Finishing (1999, 2000), together with the Breyer Medal of the Royal Australian Chemical Institute (2000) for international contributions to electrochemical science, engineering and education.

Acknowledgements

The authors have both benefited from an effective early training in physico-chemical aspects of inorganic materials. Dmitry Bavykin would highlight the contributions of Professor Savinov (Boreskov Institute of Catalysis, Novosibirsk) and Professor Parmon (Novosibirsk University), while Frank Walsh wishes to thank Drs Eddie Allen, Walford Davies, Nigel Johnson, Tom Nevell and Alan Vosper (University of Portsmouth).

This book has benefitted from research collaborations with Professor Stan Kolaczkowski, together with Drs Alexei Lapkin, Pawel Plucinski and Frank Marken (University of Bath), as well as Drs Barbara Cressey, Mark Light, Tom Markvart, Marina Carravetta and Alex Kulak (University of Southampton); the authors are also grateful to Professors John Owen and Phil Bartlett (University of Southampton) for their discussions. Some of the facilities for the characterisation of nanomaterials have been made available by Professor Mark Weller (University of Southampton). Colleagues in the national Centre for Advanced Tribology at Southampton, such as Professor Robert Wood with Drs Julian Wharton, Shuncai Wang, J. Bello and John Low, have collaborated in projects involving composite electroplated coatings incorporating titanium oxide particles. We also wish to thank Professor Gordon G. Wallace and Dr Peter Innis (Australian Research Council Centre of Excellence for Electromaterials Science at the University of Wollongong) who hosted symposia and workshops on nanomaterials in the period 2004–2009. Good laboratory facilities are essential to dyed-in-the-wool experimentalists such as the authors and provision of such facilities by the Head of the School of Engineering Sciences (Professors Geraint Price, then Mark Spearing) and the School of Chemistry (Professors Jeremy Kilburn, then John Evans) is gratefully acknowledged.

The authors wish to thank the following for financial support: the Department of Chemical Engineering at the University of Bath, the School of Engineering Sciences at the University of Southampton, the Royal Society, NATO, and EPSRC.

They gratefully acknowledge the journals Advanced Materials and the European Journal of Inorganic Chemistry for the opportunity to publish the review papers: D. V. Bavykin, J. M. Friedrich and F. C. Walsh, "Protonated Titanates and TiO$_2$ Nanostructured Materials: Synthesis, Properties and Applications," *Advanced Materials*, 2006, **18**, 2807–2824 and D. V. Bavykin and F. C. Walsh, "Elongated Titanate Structures and their Applications," *European Journal of Inorganic Chemistry*, 2009, **8**, 967–1117. The authors are also grateful to several journals for their kind permission to reprint figures (as acknowledged in the figure captions).

The authors hope that this book will provide a focussed, timely and convenient treatment of titanium oxide nanostructures in order to stimulate research work in the exciting field of inorganic nanomaterials.

Abbreviations

AAO	Anodic aluminium oxide
AFM	Atomic force microscopy
ALD	Atomic layer deposition
BET	Brunauer–Emmett–Teller (algorithm)
BJH	Barret–Joyner–Halenda (algorithm)
CTAB	Cetyltrimethylammonium bromide
CVD	Chemical vapour deposition
DFT	Density function theory
DOS	Density of states
DSC	Differential scanning calorimetry
DSSC	Dye sensitized solar cell
EBT	Eriochrome Black T (dye)
EPD	Electrophoretic deposition
ESR	Electron spin resonance
EXAFS	Extended X-ray absorption fine structure (spectroscopy)
FTIR	Fourier-transform infrared (spectroscopy)
FWHM	Full width at half the maximum height
Hap	Hydroxyapatite
HRTEM	High-resolution transmission electron microscopy
H-TiNT	Protonated titanate nanotubes
H-TiNF	Protonated titanate nanofibres
IUPAC	International Union of Pure and Applied Chemistry
MAS NMR	Magic angle spinning nuclear magnetic resonance (spectroscopy)
MB	Methylene Blue (dye)
MWCN	Multi-walled carbon nanotubes
NT	Nanotubes
NF	Nanofibres
NR	Nanorods
NP	Nanoparticles
P-25	P-25 nanoparticulate TiO_2 (Degussa)
PANI	Polyaniline
PEG	Poly(ethylene glycol)
PL	Photoluminescence
PPy	Polypyrrole

PTFE	Polytetrafluoroethylene
PVP	Polyvinylpyrrolidone
SAED	Small angle electron diffraction
SCE	Saturated calomel electrode
SEM	Scanning electron microscopy
SETOV	Single-electron-trapped oxygen vacancies
STM	Scanning tunnelling microscopy
SWCN	Single-walled carbon nanotubes
TEM	Transmission electron microscopy
TEOS	Tetraethylorthosilicate
TBAOH	Tetrabutylammonium hydroxide
TGA	Thermogravimetric analysis
THF	Tetrahydrofuran
TiNF	Titanate nanofibres
TiNT	Titanate nanotubes
TOF	Turn-over frequency
UV/VIS	Ultraviolet-visible (spectroscopy)
XANES	X-ray absorption near-edge structure (spectroscopy)
XPS	X-ray photon spectroscopy
XRD	X-ray diffraction

List of Symbols

Symbol	Meaning	Units
a'	Langmuir constant (amount of dye)	$mol(dye)\ mol(TiO_2)^{-1}$
a, b, c	Crystallographic axes	-
a_s	Amount of dye forming a monolayer on the surface	$mol(dye)\ mol(TiO_2)^{-1}$
A	Surface area	m^2
A_o	Infinity concentration of protons	$mol\ dm^{-3}$
A'	Proportionality coefficient in Equation (3.13)	-
C	Concentration of adsorbate in solution	$mol\ dm^{-3}$
d	Diameter of nanotube	m
D_{Li+}	Diffusion coefficient of lithium ions	$m^2\ s^{-1}$
E	Energy	eV
E_G	Energy bandgap of a semiconductor	eV
E_G^{1D}	Bandgap of a 1-D structure	eV
E_G^{2D}	Bandgap of a 2-D structure	eV
$E_n(0)$	Band edge	eV
$E_{2D}^{\pm}$	Energy spectrum of a TiO_2	eV
E_n	Energy of sub-bands	eV
$E_{n,1,D}$	Energy of quasi 1-D sub-bands	eV
$E_{n,1,D}^{+}$	Electronic band structure of a TiO_2 nanotube	eV
ΔE_G	Change in energy gap	eV
E_{layer}	Excess energy of a nanolayer	J
G_{2D}	Energy density of states for (2-D) nanosheets	eV^{-1}
G_{1D}	Energy density of states for (1-D) nanotubes	eV^{-1}
$G_{n,1,D}$	Density of states in each sub-band for the quasi 1-D case	eV^{-1}
H	Thickness of nanotube wall	m
h_P	Planck's constant	eV s

H_m	Bulk latent heat of fusion	J mol^{-1}
$\hbar_P$	Reduced Planck's constant	eV s
I_{sc}	Short circuit current	A
k	Spring constant	N m^{-1}
$k_\perp$	Wave vector along the circumference of a nanotube	-
$k_\parallel$	Wave vector along the axis of a nanotube	-
k_x	Wave vector along the x axis of a nanosheet	-
k_y	Wave vector along the y axis of a nanosheet	-
k_{H+}	rate constant of proton intercalation	s^{-1}
k_{Li}	rate constant of lithium ion intercalation	s^{-1}
K_L	Constant of adsorption	dm^{-3} mol
K_{sp}	Solubility product of TiO_2	(mol dm^{-3})^n
$K_{sp}^{\,b}$	Solubility product of bulk TiO_2 with a flat surface	(mol dm^{-3})^n
L	Length of an imbalanced layer in a nanosheet	m
L	Film thickness	m
L_{tube}	Length of a nanotube	m
m_e	Mass of electrons	kg
m_h	Mass of holes	kg
m_0	Rest mass of electrons	kg
n	Number of elements in each layer	-
n_i	Number of balls in the first layer	-
n_{TiO_2}	Amount of TiO_2	mol
N	Number of particles	-
p_0	Saturation vapour pressure	kPa
p/p_0	Relative saturated vapour pressure	-
r_{ext}	External radius of tube	m
r_{int}	Internal radius of tube	m
r_{curv}	Radius of curvature of a nanosheet	m
R	Boltzmann constant	J K^{-1} atom^{-1}
S	Specific surface area of a nanotube	m^2 g^{-1}
S_{ext}	External surface area of a nanotube	m^2 g^{-1}
S_{int}	Internal surface area of a nanotube	m^2 g^{-1}
T	Temperature	K
T'	Transmittance	%
T_m	Melting temperature	K
$T_m^{\,b}$	Melting temperature of an infinite bulk crystal with a flat surface	K
V_{int}	Internal pore volume	m^3
V_m	Molar volume of a liquid	m^3 mol^{-1}

V_{oc}	Open circuit potential	V
V_{tube}	Volume of a nanotube	m^3
W	Total interaction energy between two nanosheets	J
X_i	Molar fraction of component i in solution	-
Δx	Imbalance in the nanotube width	m
Δy	Curving deformation of a layer	m

Greek

α	Proportionality constant (elasticity)	$J\ m^2$
α'	Nanotube absorption coefficient	m^{-1}
β	Proportionality constant (surface tension imbalance)	$J\ m$
ε	Coefficient characterising the strength of interactions	eV
2θ	Scattering angle in XRD	deg
ϕ_1	Angle of first layer	deg
ϕ_2	Angle of second layer	deg
μ_i	Chemical potential of species i	$J\ mol^{-1}$
$\mu_{TiO_2}^b$	Chemical potential of TiO_2 without an interfacial boundary	$J\ mol^{-1}$
ν_i	Stochiometric coefficient of species i	-
ρ_s	Density of a solid material	$kg\ m^{-3}$
σ	Excess surface energy per unit area	$J\ m^{-2}$
σ_{g-l}	Interfacial tension at the liquid–gas interface	$J\ m^{-2}$

CHAPTER 1

Introduction and Scope

1.1 The History of Nanomaterials

The term *"nanostructured materials"* refers to solids which "have an internal or surface *structure* at the nanoscale. Based on the nanometre (nm), the nanoscale, exists specifically between 1 and 100 nm".[1] Although the concept of nanomaterials is relatively new, such materials have been unwittingly used for centuries. One example is colloidal gold nanoparticles dispersed in the soda glass of the famous Lycurgus Cup[2] which dates back to the fifth century and is currently displayed in the British Museum. The cup has a green appearance in reflection and red/purple colour in transmitted light. The apparent dichroism is due to the interaction of light with gold–silver and copper nanoparticles embedded into a soda glass matrix. Another example of a nanostructure is the legendary Damascus swords which contain[3] carbon nanotubes and cementite nanofibres incorporated into a steel matrix. It is possible that the unusual combination of hardness and ductility of the composite, which provides an impressive mechanical strength, flexibility and sharpness to the swords, is due to the embedded carbon nanostructures.[4] Naturally occurring clays used from the early days of pottery can also be considered as nanostructured ceramics.

More recent uses of nanostructured materials include: classical, silver image photography, which uses photosensitive nanocrystals of silver chloride; catalysis, which utilises high surface area metal nanoparticles; and painting, which uses various pigments consisting of metal or semiconductor pigment nanoparticles.[1]

Over the last two decades, improvements in technology have allowed the synthesis and manipulation of materials on the nanometre scale, resulting in an exponential growth of research activities devoted to nanoscience and nanotechnology. It is now appreciated that the physico-chemical properties of nanomaterials can be significantly different to those of bulk materials, which opens up opportunities for the development of materials with unusual or tailored properties. This has stimulated the search for methods of controlling the

RSC Nanoscience & Nanotechnology No. 12
Titanate and Titania Nanotubes: Synthesis, Properties and Applications
By Dmitry V. Bavykin and Frank C. Walsh
© Dmitry V. Bavykin and Frank C. Walsh 2010
Published by the Royal Society of Chemistry, www.rsc.org

size, shape, crystal structure and surface properties to tailor nanomaterials to a particular application.

Today, nanostructured materials are available in a wide variety of shapes including symmetrical spheres and polyhedrons, cylindrical tubes and fibres, or random and regular pores in solids. This book focuses on the *elongated* shapes of nanostructured materials, which can be defined as shapes with an aspect ratio greater than 10. The aspect ratio of the shape can be determined as the ratio between the two characteristic dimensions of a structure (*e.g.* the ratio of nanotube length to diameter).

Table 1.1 and Figure 1.1 show some examples of natural and artificial elongated nano-, micro- and macrostructures, which are prevalent in our lives. The range of the characteristic sizes and aspect ratios of these materials can cover several orders of magnitude. The composition of these structures is also very diverse.

The chain of single atoms shown at the bottom of Figure 1.1 can be considered as the tiniest possible nanostructure. Such nanostructures (*e.g.* phenylene–acetylene oligomers)[5] have recently attracted attention as possible candidates for molecular wires for use in electronic applications. Short DNA oligomers are also prospective materials for tailoring molecular nanowires, due to their versatile chemistry which facilitates functionalization and the existence of technology for sequential DNA synthesis, allowing control over the structure of biomolecules.[6]

The large class of elongated nanostructures with relatively small aspect ratios and a characteristic diameter in the range of sub to several nanometres, is represented by the elongated shape nanocrystals of semiconductor materials,[7] which have evolved from the quantum dots so actively studied over the previous decade.[8]

In comparison to nanocrystals, single-walled carbon nanotubes (SWCN) have a much higher aspect ratio and a similar range of diameters. Multi-walled carbon nanotubes (MWCN), however, are characterised by larger diameters and also very large aspect ratios (see Figure 1.1). The history of the discovery of carbon nanotubes is still the subject of debate.[9] The first TEM image of carbon nanotubes was reported in 1952.[10] At this time, research was focussed on the prevention of nanotube growth in the coal and steel industry and in the coolant channels of nuclear reactors. In 1991, carbon nanotubes were rediscovered by Iijima,[11] followed by a massive interest in these structures from the scientific community. In turn, this has led towards the discovery of many other materials having a nanotubular morphology.

1.1.1 The Importance of TiO$_2$ and Titanate Nanomaterials

Despite the relatively high abundance of titanium in nature and the low toxicity of most of its inorganic compounds, the metallurgical cost of extracting titanium metal is high due to the complexity of the traditional Kroll molten salt extraction process. The original demand from aerospace and rocket jet

Table 1.1 Examples of important elongated structures. Values are approximate. (SWCN: single-walled carbon nanotubes; MWCN: multi-walled carbon nanotubes).

Name	Diameter, d	Length, L	Aspect ratio $= L/d$	Composition	Comment
Tobacco mosaic virus	18 nm	50–300 nm	3–16	DNA, protein	Well characterized model system
Chromosome (human) in metaphase	700 nm	5–10 μm	7–30	DNA, protein	A compact form of DNA
SWCN	1–4 nm	100 nm–10 mm	3×10^5	Carbon	Popular objects for the study of fundamental science
MWCN	30 nm	100 nm–10 mm	10^5	Carbon	Popular for study of fundamental science
TiNT	7–20 nm	50–2000 nm	5–200	Titanic acid	Focus of this book
TiNF	50–200 nm	1–10 μm	5–200	Titanic acid	Focus of this book
Human hair	20–200 μm	0.01–1 m	10^4	Protein	everyday object
Synthetic fibres (nylon)	>10 μm	Unlimited	Unlimited	Polymers	Used in the textile industry
Carbon fibres	5–10 μm	Unlimited	Unlimited	Carbon	Used as porous electrodes
Optical fibres	0.6 mm	1–2 km	3×10^6	Glass	Used in internet communication
Bamboo tree	1–10 cm	Up to 100 m	10^3	Cellulose	Natural plant with a large aspect ratio

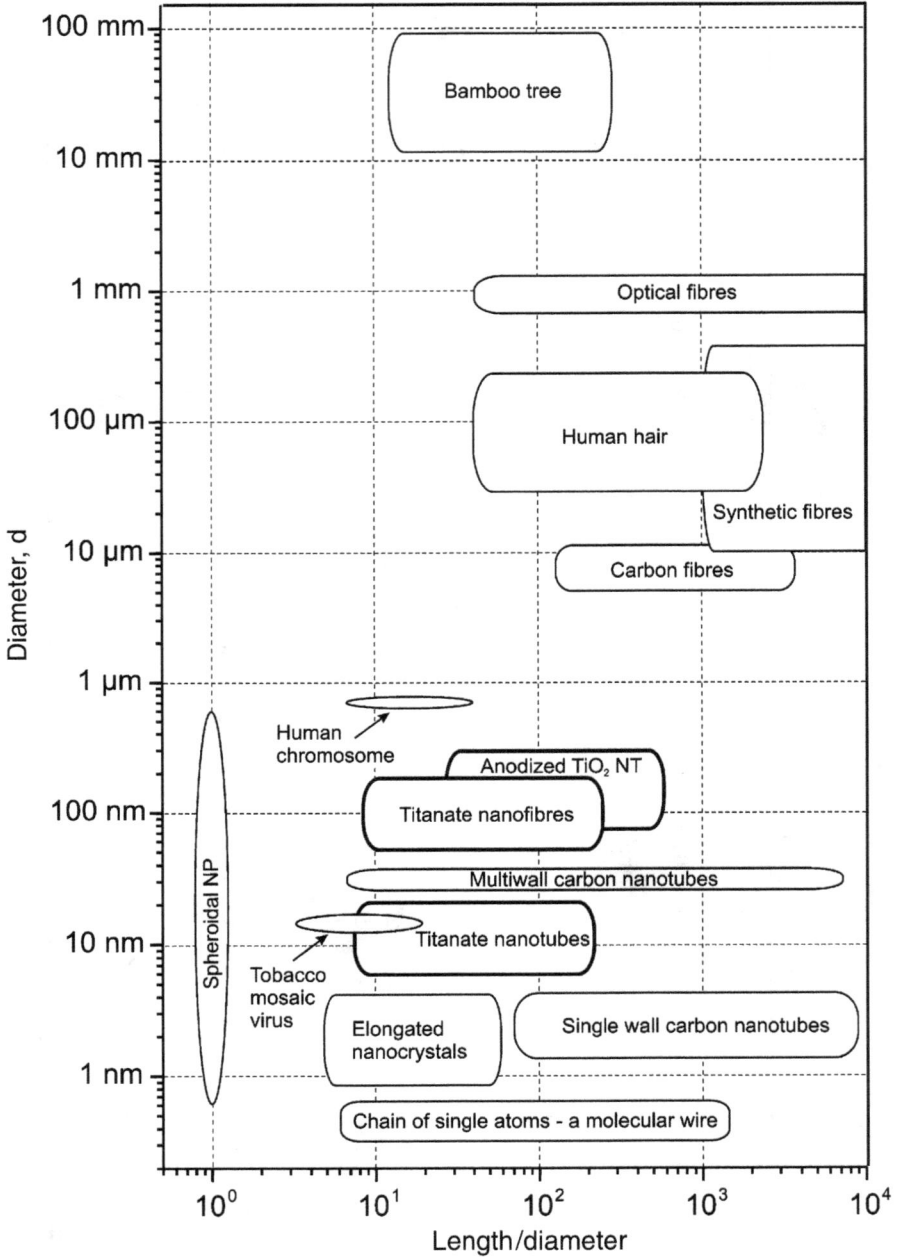

Figure 1.1   Abundance of elongated structures.

industries for the lightweight, high melting temperature metal in the late 1940s, stimulated improvements in the Kroll extraction process and initiated large-scale titanium production. In the late 1960s, approximately 80% of the titanium produced was used in the aerospace industry.[12] Further reductions in the manufacturing cost of titanium have also stimulated the use of titanium compounds. Titanium dioxide has long been used as a white pigment in paints and polymers. Following the discovery of photocatalytic water splitting using TiO_2 under UV light[13] in the late 1970s, a new era of TiO_2-based materials has emerged.[14]

Following developments in nanotechnology, similar trends have occurred in the synthesis of nanostructured titanium dioxide and titanate materials. Initially, many of the nanostructured TiO_2 materials, produced mainly by a variety of sol–gel techniques, consisted of spheroidal particles whose size varied over a wide range down to a few nanometres. The most promising applications of such TiO_2 nanomaterials were photocatalysis, dye sensitised photovoltaic cells and sensors.[14]

In 1998, Kasuga and colleagues[15] discovered the alkaline hydrothermal route for the synthesis of titanium oxide nanostructures having a tubular shape. The search for nanotubular materials was inspired by the rediscovery of carbon nanotubes in 1991.[11] Studies of their elegant structure and unusual physico-chemical behaviour have significantly improved our fundamental understanding of nanostructures. In contrast to carbon nanostructures, titanate and titanium dioxide nanotubes are readily synthesised using simple chemical (*e.g.* hydrothermal) methods using low cost materials.

Following the discovery of titanate nanotubes, many efforts have been made to: (*a*) understand the mechanism of nanotube formation, (*b*) improve the method of synthesis, and (*c*) thoroughly study the properties of nanotubes. Other elongated morphologies of nanostructured titanates, including nano-rods, nanofibres and nanosheets, have also been found. Many data have been collected and presented in recent reviews.[16–19]

Under alkaline hydrothermal conditions, the formation of titanate nano-tubes occurs spontaneously and is characterised by a wide distribution of morphological parameters, with a random orientation of nanotubes. An alternative method, which facilitates a structured array of nanotubes with a narrower distribution of morphological parameters, is anodising. Anodic synthesis was initially developed for the preparation of aluminium oxide nanotubes, and later adapted for nanotubular TiO_2 arrays. The method includes anodic oxidation of titanium metal in an electrolyte, usually containing fluoride ions. Control of the fabrication conditions enables a variation in the internal diameter of such nanotubes from 20 to 250 nm, with a wall thickness from 5 to 35 nm, and a length of up to several hundred microns.[20] Several major reviews which consider the manufacture, properties and various applications of these ordered TiO_2 nanotubular coatings have recently been published.[20–22]

The third general method for the preparation of elongated TiO_2 nanos-tructures is template-assisted sol–gel synthesis. This versatile (but sometimes

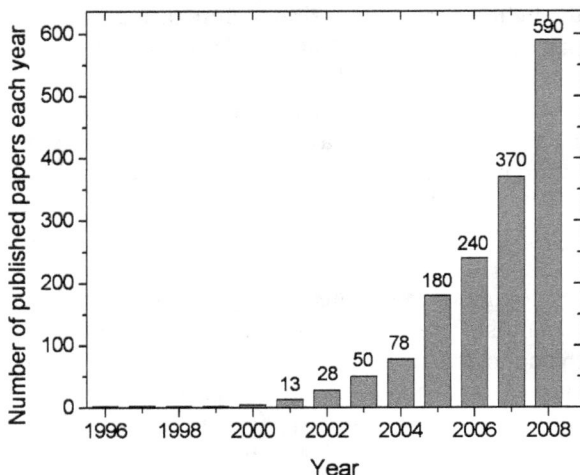

Figure 1.2   The number of papers related to TiO_2 and titanate nanotubes as a function of the year of publication. (Data were collected from the ICI Web of Science® database using "TiO_2 and nanotube*" as keywords).

expensive) technique is reviewed elsewhere.[23] The synthesis of TiO_2 and titanate nanotubes is considered in Chapter 2.

Since the discovery of TiO_2 nanotubes, the amount of published material relating to this subject is growing exponentially year by year (see Figure 1.2), indicating the great interest. The pool of published work in the area of elongated titanates and TiO_2 can be classified according to several themes: (*a*) the improvement in the methods of nanostructure formation in order to better control morphology and lower manufacturing costs, including mechanistic studies, (*b*) the exploration of the physical chemical properties of novel nanostructures, with a focus on potential applications, and (*c*) the use of elongated titanates in a wide range of applications. Since the discovery of titanate nanotubes, the amount of published work relating to the first two themes has rapidly grown (and may be approaching a steady state), whereas the third theme has appeared only recently and is experiencing a rapid growth.

1.2   Classification of the Structure of Nanomaterials

The field of nanoscience is relatively young and a number of new terms have appeared, some of which are inconsistent. It is unfortunate that the definition of various morphological forms of the nanomaterials has not taken place in a careful fashion, which can result in some confusion over their use. In this book, the following terms for various shapes of nanostructured TiO_2 and titanate will be used (see Figure 1.3). The proposed morphologies are consistent with recently suggested classifications.[24]

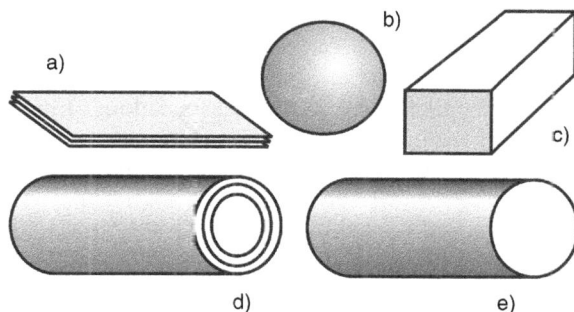

Figure 1.3 Idealised morphologies of elongated titanate and TiO$_2$ nanostructures: a) sheets, b) spheroids c) rectangular section fibres, d) multiple wall nanotubes, and e) circular section rods.

Nanotubes (or *nanoscrolls*) shown in Figure 1.3d are long cylinders having a hollow cavity positioned at their centre and lying along their length. The aspect ratio (*i.e.* the length divided by the diameter) of nanotubes is usually > 10, and can achieve several thousand. The walls of titanate nanotubes are always multilayered and the number of layers varies from 2 to 10. Structurally, nanotubes can be scrolled, "onion" or concentric in type. Sometimes, the single nanotube has a different number of layers in the two different walls in the axial cross sections of the tube obtained by TEM imaging. Nanotubes are usually straight with a relatively constant diameter. However, small amounts of tubes with a variable internal diameter and closed ends are also found. TiO$_2$ nanotubes produced by anodic oxidation of titanium, always have one open end and another end which is closed.

Titanate nanotubes are usually produced by folding *nanosheets*, as indicated in Figure 1.3a. There are two types of nanosheets: *single layer* nanosheets, which are isolated (100) planes of titanates, or *multilayer* nanosheets, which are several conjugated (100) planes of titanates. Both types of nanosheets are very thin and could be found in both planar or curved shapes. The typical dimensions of nanosheets are < 10 nm thickness, and > 100 nm height and width. Nanosheets are usually observed in the early stage of preparation of nanotubes or as a small impurity in the final product obtained *via* the alkaline hydrothermal route.

Nanowires or *nanorods*, seen in Figure 1.3e, are long, solid cylinders with a circular base, nanowires being longer than nanorods.[24] Both morphologies do not usually have internal layered structures and have a similar aspect ratio to nanotubes. Nanowires can often be found in samples of nanotubes annealed at temperatures above 400 °C (see Chapter 4 for details).

Long, solid, parallel-piped titanates are termed *nanoribbons*, *nanobelts* or *nanofibres* in the literature (see Figure 1.3c). These structures tend to have a good crystallinity, and the relationship between the length of the edges corresponding to each crystallographic axis is usually in the order $l_{001} \gg l_{100} >$

l_{010} (ref. 25). The length of the nanofibres (l_{001}) can be several tens of microns, while the width of nanofibres (l_{001} or l_{010}) is typically in the range 10–100 nm. The aspect ratio can be as large as several thousand. Nanofibres, which are usually produced during alkaline hydrothermal reactions at high temperatures, can be found in straight, as well as curved forms.

During hydrothermal treatment, individual morphological forms of titanates tend to agglomerate into secondary particles. The resulting textures include nanotubular bundles,[26] split nanofibres and hierarchical linked nanofibres[27] *etc.* Unfortunately, there are few reported systematic studies which allow for a comprehensive treatment of the reasons for producing a given texture. There is an even less systematic terminology describing these secondary agglomerates.

1.3　Synthesis of Important Elongated Nanomaterials

In this section, non-templated synthetic methods for the preparation of several nanotubular and nanofibrous materials, with nanostructures based on compounds other than TiO_2, are reviewed. The list of examples presented here is not comprehensive and covers mostly hydrothermal wet chemical methods. Particular interest is focussed on examples of spontaneous formation of multilayered walls nanotubes. Methods involving template synthesis are excluded, since they are relatively general and can be a subject of a separate book. Brief descriptions of synthetic procedures are provided. The selected nanostructures are believed to have potential applications in various areas of technology and nanotechnology.

1.3.1　Metal Oxide Nanotubes

Aluminium oxide nanotubes. The anodic oxidation of aluminium in acidic electrolytes, resulting in the formation of a porous film of anodic aluminium oxide (AAO) consisting of nanotubes, was discovered in a pre-nanoscience era.[28] In a typical method, pure aluminium is anodised in the presence of oxalic acid ($0.3 \, mol \, dm^{-3}$), at a constant cell voltage of 40 V. A long period of anodisation can improve the regularity of the tube arrangement, towards a hexagonal array of nanotubes.[29] AAO films are currently widely used as a template for the preparation of other materials with a nanostructured morphology.

Aluminium oxide nanorods are usually produced by the hydrolysis of aluminium chloride during electrospinning[30] or in hydrothermal conditions.[31] In a typical hydrothermal procedure, 0.724 g of pure $AlCl_3 \cdot 6H_2O$ is dissolved in 30 cm^3 of water at room temperature, followed by the slow addition of 15 cm^3 of aqueous $Na_2B_4O_7 \cdot 10H_2O$ ($0.1 \, mol \, dm^{-3}$) with vigorous stirring. The transparent solution is then transferred into a PTFE-lined autoclave and hydrothermally treated at 200 °C for 24 h. The chemical composition of the nanorod bundles produced is similar to that of boehmite (γ-AlOOH).

Barium titanate nanotubes ($BaTiO_3$ NT) can be produced *via* a hydrothermal reaction of a TiO_2 nanotube array with excess of aqueous $Ba(OH)_2$ ($0.05\,mol\,dm^{-3}$) at $150\,°C$ for 2 h.[32] The morphology of barium titanate nanotubes is similar to that of the original TiO_2 nanotubes.

Bismuth oxyhalides nanotubes ($Bi_{24}O_{31}Br_{10}$ NT) can be produced by the hydrolysis of $Bi(NO_3)_3$ (0.5 g) in the presence of CTAB (0.5 g) with the addition of NaOH to a pH of 10, followed by hydrothermal treatment at $100–120\,°C$ for 2–4 h.[33] Bismuth oxyhalide nanotubes are characterised by a multilayer wall structure, with a diameter of 3–8 nm and a length of between 2 and 5 μm. An increase in hydrothermal temperature or the synthesis time, results in the formation of nanofibres rather than nanotubes.

Cobalt oxide nanotubes (CoO and Co_3O_4 NT) can be synthesised by the slow addition of $10\,cm^3$ of a $NH_3 \cdot H_2O$ solution ($0.1\,mol\,dm^{-3}$) to $10\,cm^3$ of a $Co(NO_3)_2$ solution ($0.025\,mol\,dm^{-3}$). After stirring for 15 min, the precipitate is washed with water followed by filtration. The wet precipitate is mixed with a water–methanol mixture (1 : 1 v/v) and 0.3 g $NaNO_3$ is added for the preparation of Co_3O_4 nanotubes (otherwise CoO nanotubes are formed), followed by hydrothermal treatment at $250\,°C$ for 24 h.[34] Both nanotubes are characterized by a multilayer wall structure with a 0.7 nm interlayer spacing. The outer diameter of nanotubes ranges from 10 to 20 nm and the nanotubes are up to several micrometres long.

Germanium oxide nanofibres (GeO_2) can be synthesised by the simple hydrothermal recrystallisation of 1 g of bulk GeO_2 in $48\,cm^3$ of distilled water at $450\,°C$ and 8.5–9.3 MPa, and stirring with a rotating speed of 100 rpm for 24 h.[35] After the hydrothermal process, the nanofibres can be collected from the internal surface of the autoclave. The nanofibres are characterised by a single crystal structure of 30–300 nm diameter, and a length > 10 μm.

Hafnium oxide nanotubes (HfO_2 NT) can be prepared by the electrochemical oxidation of hafnium in a fluoride-containing acid electrolyte. In a typical procedure, pure hafnium foil is degreased by sonicating it in acetone, isopropanol and methanol sequentially, followed by rinsing with water and drying in a nitrogen atmosphere. Electrochemical anodization takes place in H_2SO_4 ($1\,mol\,dm^{-3}$) with an aqueous NaF addition to the electrolyte (0.2 wt%), at room temperature with a cell voltage of 10 to 60 V.[36] The nanotubes obtained have an internal diameter in the range of 15 to 90 nm (depending on the voltage), and are several tenths of a micrometre in length.

Iron oxide nanotubes (α-Fe_2O_3 NT, hematite) can be synthesised by the hydrothermal treatment of a mixture of $FeCl_3$ and $NH_4H_2PO_4$ solutions. In a typical experimental procedure, $3.20\,cm^{-3}$ of an aqueous $FeCl_3$ solution ($0.5\,mol\,dm^{-3}$) and $2.88\,cm^{-3}$ of an aqueous $NH_4H_2PO_4$ solution ($0.02\,mol\,dm^{-3}$) are mixed with vigorous stirring, followed by the addition of water to give a final volume of $80\,cm^{-3}$. The mixture is then hydrothermally treated at $220\,°C$ for 48 h. The product consists almost entirely of nanotubes with outer diameters of 90–110 nm, inner diameters of 40–80 nm and lengths of 250–400 nm.[37]

Lead titanate nanotubes (PbTiO$_3$ NT) can be obtained using similar methods to those used in the preparation of barium titanates, by the hydrothermal treatment of a TiO$_2$ nanotube array with an aqueous solution of lead acetate (0.001 mol dm^{-3}) at 260 °C for 6 h.[38] Lead titanate nanotubes demonstrate good piezoelectric and ferroelectric properties.

Magnesium hydroxide nanotubes (Mg(OH)$_2$ NT) can be prepared using a two-stage method *via* an intermediate product of Mg$_{10}$(OH)$_{18}$Cl$_2 \cdot$5H$_2$O nanofibres.[39] The hydrothermal treatment of this intermediate at 180 °C for 6 h in ethylenediamine (without stirring), results in the formation of lamellar nanotubes with an outer diameter of 80–300 nm, a wall thickness of 30–80 nm and a length of several tenths of a micrometre. A recent alternative method allows for the production of a smaller size of nanotubes.[34] According to this method, 20 cm^3 of an aqueous solution of MgCl$_2$ (0.15 mol dm^{-3}) is slowly mixed with 10 cm^3 of a NH$_3 \cdot$H$_2$O solution (5 mol dm^{-3}), with vigorous stirring at room temperature. The precipitate is thoroughly washed with water, and hydrothermally treated in 18 cm^3 of a water–methanol mixture (1 : 1 v/v) with the addition of 0.3 g NaNO$_3$ at 250 °C for 36 h. The nanotubes formed have a diameter of 10–20 nm and a length of several micrometres.

Manganese(IV) oxide nanotubes (MnO$_2$ NT) can be produced using several methods, each of which results in the formation of nanotubes with a specific morphology and crystal structure. Solid MnO$_2$ can exist in α, β, δ and γ crystallographic structures. The most stable is β-MnO$_2$ which can be obtained with a nanotubular microstructure using the following method. In a typical process, 4.0 mmol of MnSO$_4 \cdot$H$_2$O is dissolved in 10 cm^{-3} of distilled water, and 4.5 mmol of PVP (K30, polymerization degree 360) is added slowly with vigorous stirring. When the solution becomes transparent, 8 cm^{-3} of an aqueous solution containing 8.0 mmol of NaClO$_3$ is added with continuous stirring. The resulting transparent solution is hydrothermally treated at 160 °C for 10 h. The β-MnO$_2$ nanotubes obtained are usually cylindrical in shape, with a diameter of 200–500 nm and a length of 1–6 μm.[40]

δ-MnO$_2$ nanotubes can be obtained *via* the disproportion–hydrolysis of α-NaMnO$_2$. In a typical synthesis, 0.3 g of NaMnO$_2$ is dispersed into 30 cm^{-3} of diluted water, followed by hydrothermal treatment at 120–140 °C for about 4 days.[41] The nanotubes are characterized by a multilayer wall structure (interlayer distance of 0.7 nm), with diameters of 10–20 nm and lengths of several μm (see Figure 1.4a). Similar multilayered MnO$_2$ nanotubes can also be produced using a multistage approach *via* the formation of single-layer MnO$_2$ nanosheets.[42]

Nickel hydroxide nanotubes (Ni(OH)$_2$ NT) can be obtained by the hydrothermal treatment of the precipitate formed during the slow mixing of 10 cm^3 of NH$_3 \cdot$H$_2$O solution (0.2 mol dm^{-3}) with 10 cm^3 of Ni(NO$_3$)$_2$ solution (0.025 mol dm^{-3}). A thorough wash of the precipitate with water, followed by hydrothermal treatment in 18 cm^3 of a water–methanol mixture (1 : 1 v/v) with addition of 0.3 g of NaNO$_3$ at 250 °C for 24 h, results in the formation of multilayer wall nanotubes with 0.7 nm interlayer spacing, 10–20 nm diameter and several micrometres length.[34] The lamellar wall structure of nanotubes is

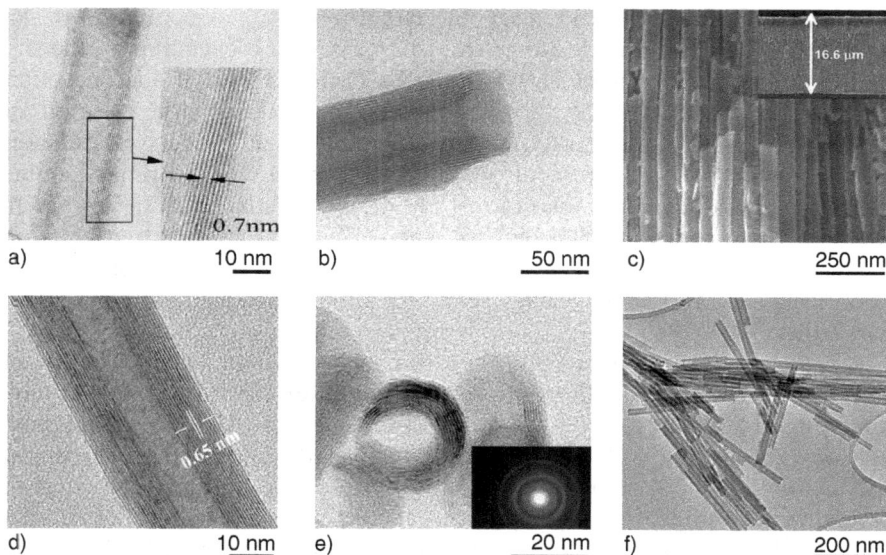

Figure 1.4 Electron microscopy images of MnO_2: a) VO_x, b) ZrO_2 c) WS_2, d) $Ni_3Si_2O_5(OH)_4$ and e) $Mg_3Si_2O_5(OH)_4$ nanotubes. (Images a), b), c), d) and e) were kindly taken with permission from ref. 41, 47, 50, 55, 61 respectively).

similar to that of $Ni(SO_4)_{0.3}(OH)_{1.4}$ *nanofibres* obtained under similar hydrothermal conditions.[43]

Niobate nanotubes (e.g. $K_4Nb_6O_{17}$ NT) can be fabricated using a multistage approach similar to that used for MnO_2 nanotubes. In the first stage, potassium ions from bulk multilayered $K_4Nb_6O_{17}$ are ion-exchanged to protons using strong inorganic acids (*e.g.* HCl). In the second stage, niobate single-layer nanosheets are exfoliated from the bulk crystal using TBAOH, which intercalates between the layers. In the third stage, the colloidal solution of single layer nanosheets is coagulated by the addition of KCl or NaCl solution forming nanotubes of alkaline metal niobates.[42,44] The nanotubes that are obtained are characterised by a multilayered structure, with an external diameter ranging from 15 to 30 nm and a length of 0.1 to 1 µm.

Vanadium(V) oxide nanotubes (VO_x NT, where $x \approx 2.45$) were originally synthesised using the hydrolysis of vanadium(v) alkoxide in the presence of an amine, followed by hydrothermal treatment. In a typical reaction procedure, a solution of vanadium(v) triisopropoxide (15.75 mmol) and hexadecylamine (7.87 mmol) in a molar ratio of 2 : 1 in ethanol (5 cm^3) is stirred in an inert atmosphere for 1 hour, followed by the addition of 15 cm^3 of water. After a few hours, a yellow, lamellar-structured composite of surfactant and a hydrolyzed vanadium oxide component are formed. The hydrothermal reaction of the mixture at 180 °C for about a week, leads to the formation of black, isolated or star-like grown-together nanotubes with open ends as the main product.[45]

VO_x–nanotubes are characterized by multilayer wall structures with an interlayer spacing varying in range from 1.7 to 3.8 nm (depending on the size of the amine additive used in synthesis) as seen in Figure 1.4b. The outer diameter of the nanotubes ranges from 15 to 150 nm, with a length of 0.5–15 μm.[46,47] A more recent, alternative method of nanotube synthesis uses another precursor. A suspension of V_2O_5 (15 mmol) and a primary amine $C_nH_{2n+1}NH_2$ ($11 < n < 20$; molar ratio 1 : 1) in 5 cm^3 of ethanol is stirred for 2 h, followed by the addition of 15 cm^3 of water. The mixture is stirred at room temperature for 48 h. The hydrothermal treatment of the resulting composite in an autoclave at 180 °C for 7 days generates a black product containing VO_x nanotubes.[48]

Zinc oxide nanotubes (ZnO NT) can be prepared by hydrolysis of $Zn(NH_3)_4^{2+}$ complexes under hydrothermal conditions. In a typical procedure, a mixture of aqueous $Zn(NH_3)_4^{2+}$ solution (adjusted to pH 12 with ammonia) and ethanol (1 : 11 v/v) is hydrothermally treated with constant stirring at 180 °C for 13 h.[49] The resultant white precipitate can be collected, then washed several times with absolute ethanol and distilled water. The yield of tubular ZnO is about 20% polycrystalline nanotubes of approximately 450 nm in diameter and 4 μm in length.

Zirconium oxide nanotube (ZrO$_2$ NT) arrays can be fabricated by anodic processing of pure zirconium metal in a fluoride-containing electrolyte. According to one method, a degreased and clean, flat zirconium metal electrode is electrochemically oxidized in an aqueous electrolyte containing $(NH_4)_2SO_4$ solution (1 mol dm^{-3}) and 0.5 wt% NH_4F at room temperature, at a cell voltage of 20 V.[50] ZrO$_2$ nanotubes are characterised by a 50 nm internal diameter and are approximately 17 μm in length (see Figure 1.4c).

1.3.2   Metal Chalcogenide Nanotubes

Chalcogenides of transition metals having a multi-walled nanotubular morphology have been actively studied during recent decades. The methods of nanotube preparation include arc discharge; laser ablation; sublimation; gas-phase reduction with H_2S or H_2Se; pyrolysis of $(NH_4)_2MX_4$ (where M = Mo or W; X = S or Se); and hydrothermal reactions.[51] Although gas-phase redox synthesis is the most prominent technique used, in this book we mainly review hydrothermal methods. During hydrothermal sulfurisation, the following molecules are used as a source of sulfur atoms: H_2S, $(NH_2)_2CS$ and KSCN.

Bismuth(III) selenide nanotubes (Bi$_2$Se$_3$ NT) can be prepared by the hydrothermal reduction of H_2SeO_3 in the presence of Bi^{3+} ions. In a typical procedure, 7.5 cm^3 of aqueous H_2SeO_3 (1 mol dm^{-3}) is mixed with 5.0 mmol of $BiCl_3$ under ultrasonic agitation for about 30 min, followed by the addition of 4 cm^3 of hydrazine hydrate ($N_2H_4 \cdot H_2O$). The mixture is hydrothermally treated at 200 °C for 24 h.[52] The black-coloured product obtained contains a certain amount of nanotubes which are impurities to other nanostructures including nanosheets and nanofibres. The nanotubes are characterised by internal diameters of 5–10 nm, lengths of 30 to 120 nm and approximate wall thicknesses of 1.3 nm.

Molybdenum(IV) polysulfide nanotubes (MoS$_8$ NT) can be hydrothermally prepared from ammonium salts. In a stainless-steel autoclave vessel, solutions of 2.8 g of $(NH_4)_2MoS_4$ in 100 mL of water and 8 g of $(NH_2OH)_2 \cdot H_2SO_4$ dissolved in 100 mL of water are mixed together, followed by the addition of gaseous H_2S to a pressure of 2000 kPa. A hydrothermal reaction is then performed at 220 °C for 30 min. A brown powder of nanotubes can be filtered off, washed with distilled water and dried in air. The hollow tubules of MoS$_8$ are of nanoscale size (50–500 nm diameter; 0.1–20 μm length).[53]

Molybdenum(IV) sulfide nanotubes (MoS$_2$) can be obtained *via* the hydrothermal reduction of MoO_3 with KSCN. In a typical preparation of MoS$_2$ nanotubes and nanorods, 5.0 mmol of molybdenum(VI) oxide (MoO_3) and 12.5 mmol of potassium thiocyanate (KSCN) are mixed with 60 cm^3 of water, and hydrothermally treated at 180 °C for 24 h. A dark powder consisting of polycrystalline nanotubes is collected and washed several times with deionized water to remove any residue of the reactants. The nanotubes are characterized by a diameter of 60–100 nm and a length of several microns.[54]

Tungsten sulfide nanotubes (WS$_2$ NT) can be synthesized *via* the gas-phase reduction of WO_3 nanorods with H_2S at elevated temperatures. WO_3 nanorods can be obtained *via* the following route. An aqueous solution containing a mixture of 1.32 g of $(NH_4)_{10}W_{12}O_{41} \cdot 7H_2O$ and 2.10 g of citric acid is heated at approximately 120 °C with constant stirring for 4–5 h until a gel is formed, and allowed to stand overnight. Hexadecyl amine (2.45 g) dissolved in ethanol is added to the gel with stirring for 10 h. The resulting mixture is hydrothermally treated at 180 °C for 7 days. The product obtained is washed with ethanol, cyclohexene, water, and finally again with ethanol, and dried at room temperature. An alumina crucible containing the WO_3 nanorods is placed in a tubular furnace and heated up to 840 °C at a rate of 5 °C min^{-1} in an Ar gas flow, and then it is switched to H_2S gas for 30 min.[55] The diameter of the nanotubes obtained varies broadly, ranging from 20 to 200 nm, and their length varies from approximately 1 to 8 μm. A large fraction of the nanotubes has open ends. The walls have a layered structure with an interlayer spacing of 0 65 nm (see Figure 1.4d).

Titanium sulfide nanotubes (TiS$_2$ NT) can be obtained *via* hydrothermal re-crystallisation. In a typical procedure, anhydrous sodium sulfide (3.1 g, 40 mmol) is added to the solution of titanium(IV) chloride (3.8 g, 20 mmol) in tetrahydrofuran, THF (150 cm^3), at 40 °C with stirring. After 30 min, the liquid is cooled and the dark brown solid is filtered, and washed with THF and methanol several times. Hydrothermal treatment of the powder in THF at 200 °C for 6 h, results in the re-crystallization of TiS$_2$ forming nanotubular-shaped single crystals. During synthesis, care should be taken to avoid any impurities in the water.[56] The resulting nanotubes have the following approximate dimensions: a length of 2 μm, an outer diameter of 30 nm, an inner diameter of 10 nm; and multilayered walls with an interlayer spacing of 0.57 nm.

1.3.3   Mixed Oxides, Silicates and Other Compounds as Nanotubes

The chemistry of layered mixed silicon oxide with other metal oxide compounds is rather complex. Various metal silicates are abundant in nature. Some of them occur naturally with a nanotubular morphology, which can be reproduced artificially using the hydrothermal processes similar to those occurring in the earth's crust. In the following sections, we review several examples of mixed oxide nanotubes and their syntheses.

Chrysotile nanotubes ($Mg_3Si_2O_5(OH)_4$ NT) occur naturally in the earth crust where they are probably hydrothermally produced under high pressure and temperature. Such geological conditions have inspired the preparation of synthetic chrysotile.[57] The development of the synthetic methods for chrysotile preparation began in the pre-nanotube era of the early 1920s. Since then there have been many improvements with an optimisation of reaction conditions to achieve a better control of morphology, and the ability to perform syntheses using ambient reaction conditions. One recent method of synthesis includes the following steps. Silica (SiO_2; prepared by hydrolysis of TEOS) is mixed with $Mg(OH)_2$ (prepared by hydrolysis of magnesium ethoxide) in a molar ratio of 3 : 2 (Mg : Si). A 0.3 g portion of the mixture was added to aqueous NaOH (pH = 13) and was hydrothermally treated at 200 °C for 5 days.[58] The nanotubes produced had an average outer diameter < 50 nm, with a length which could exceed 1 μm, as seen in Figure 1.4f. The walls had a multilayered structure with an interlayer spacing of 0.73 nm.

Halloysite nanotubes ($Al_2Si_2O_5(OH)_4 \cdot 2H_2O$ NT) naturally occur in mineral form and are relatively abundant on the surface of the Earth. The length of natural nanotubes varies from 1 to 15 μm with an inner diameter from 10 to 150 nm, depending on the mining location. The tube has a multilayered wall structure. There appears to be no reported laboratory synthesis of halloysite nanotubes, despite the fact that the preparation of synthetic nanotubes could provide a method for obtaining the material free from the impurities which are always present in natural minerals.

Hydrotalcite nanoscrolls ($Mg_6Al_2(CO_3)(OH)_{16} \cdot 4H_2O$) are the layered double hydroxides which can form nanotubular structures (scrolls). These hydroxides can be prepared *via* the hydrothermal hydrolysis of aluminium and magnesium salts in a water–ethanol solvent using the following procedure: water–ethanol solutions of $Mg(NO_3)_2 \cdot 6H_2O$ (1.8 mmol) and $Al(NO_3)_3 \cdot 9H_2O$ (0.6 mmol) are added to a solution of NaOH (0.0025 mol) and Na_2CO_3 (0.0025 mol) in 30 cm^3 of water with stirring, followed by hydrothermal treatment at 160 °C for 24 h.[59] The resulting nanotubes are characterized by outer diameters in the range of 50–100 nm, and lengths ranging from hundreds of nanometers to several microns.

Imogolite nanotubes ($(OH)_3Al_2O_3SiOH$ NT) are naturally occurring single-walled nanotubular aluminosilicates with an approximate inner diameter of 1.0 nm, an external diameter of 2.5 nm, and a length ranging from hundreds of nanometres to several microns. In nature, these aluminosilicates are usually found in the decomposition products of volcanic ash, in soils with a humid environment. Synthetic imogolite, however, can be prepared using following procedure: TEOS is added dropwise to an aqueous aluminum chloride solution $(2.4 \, \text{mmol} \, \text{dm}^{-3})$ to give a molar ratio of 9 : 5 (Al : Si) . The solution is then adjusted to pH 5.0 using aqueous NaOH solution $(0.1 \, \text{mol} \, \text{dm}^{-3})$ for the hydrolysis of the precursors, followed by the addition of solutions of HCl $(1.0 \, \text{mmol} \, \text{dm}^{-3})$ and ethanoic acid $(2.0 \, \text{mmol} \, \text{dm}^{-3})$. The acidified, aqueous solution is then refluxed at 98 °C for 5 days. The separation of the nanotubes from unreacted silicic acid, monomeric aluminum and NaCl is achieved by using dialysis against deionized water.[60]

Nickel(II) phyllosilicate nanotubes $(Ni_3Si_2O_5(OH)_4$ NT) are a homologous analogue of chrysotile nanotubes and can be obtained *via* a hydrothermal process. In a typical procedure, $NiCl_2 \cdot 6H_2O$ (3.21 g; 13.5 mmol) is added to a suspension of silicic acid (0.70 g; 9.0 mmol) dispersed in $200 \, \text{cm}^3$ of distilled water under a nitrogen atmosphere, followed by the slow addition of $67.5 \, \text{cm}^3$ of NaOH solution $(1.0 \, \text{mol} \, \text{dm}^{-3})$ over a period of 10 min. After 3 days stirring under N_2 at room temperature, the mixture is transferred to an autoclave and hydrothermally treated at 250 °C for 18 h.[61] The prepared nanotubes have a relatively short length of up to 200 nm. The outer and inner diameters of the tubes range from 25 to 30 nm and from 10 to 15 nm, respectively. The walls of the nanotubes have a multilayered structure with a 0.7 nm spacing between the layers, as seen in Figure 1.4e. Using a similar approach it is also possible to prepare Co–Mg phyllosilicate nanostructures.[62]

Transition metal *silicate nanotubes* (*e.g.* copper silicates; $CuSiO_3 \cdot 2H_2O$) can be obtained using the following procedure:[63] $Cu(NO_3)_2 \cdot 3H_2O$ (0.5 g) is dissolved in a mixture of distilled water (5 mL) and ethanol (20 mL), into which ammonia solution is added to form a clear $[Cu(NH_3)_4]^{2+}$ solution. Then Na_2SiO_3 (5 mL; $0.5 \, \text{mol} \, \text{dm}^{-3}$) is added to form a light-blue precipitate, which, following hydrothermal treatment at 180–200 °C for two days, results in the formation of nanotubes.

Silica nanotubes $(SiO_2$ NT) can be produced from kaolin mineral $(Al_2Si_2O_5(OH)_4)$ *via* the intercalation of surfactant between layers on the precursor followed by delamination and hydrothermal transformation to nanotubes. According to this procedure, the white colored kaolin powder is obtained by calcination of the original kaolin at 750 °C for 4 h in air. The soluble salts are then removed by washing the material with distilled water and drying it at 100 °C. The dried kaolin (5 g) is saturated with $20 \, \text{cm}^3$ of aqueous CTAB $(0.1 \, \text{mol} \, \text{dm}^{-3})$ for 10 h at room temperature, followed by the addition of $2 \, \text{cm}^3$ of concentrated sulfuric acid and stirred for 20 h.[64] The suspension which is obtained is heated to 120 °C for 24 h. The nanotubular structures produced are characterized by an open ended tubular morphology, with an inner diameter of about 50 nm, an outer diameter of 80 nm and a length of less than 1 μm on average.

Bismuth metal nanotubes can be obtained by reduction of bismuth nitrate with hydrazine in hydrothermal conditions. During this procedure, 0.01 mol of $Bi(NO_3)_3$ is mixed with 0.02 mol of aqueous N_2H_4 in water at room temperature, forming an insoluble precipitate. After adjusting pH to within the range of

12–12.5 with an aqueous solution of NH_3, the mixture is hydrothermally treated at 120 °C for 12 h, and then washed with HCl. The diameter of the nanotubes produced is approximately 5 nm, and the length is in the range of 0.5 to 5 μm.[65]

In conclusion, the list of elongated inorganic nanostructures is constantly growing due to the interest in new nanomaterials with nanotubular and nanofibrous morphology.[66] At present, the search for new materials tends to apply methods of *trial and error*, due to the lack of a general theory regarding nanostructure growth which could allow for the prediction of the synthesis conditions required for new nanomaterials. However, current research on the mechanisms of nanostructure formation (see Chapter 2) is providing a route for the development of more consistent theories.

1.4 Techniques and Instruments Used to Study Nanomaterials

The current boom in nanoscience and nanotechnology is to some extent a result of the recent advances in the methods of investigating and manipulating small objects. Such developments were mostly driven by the improvement in the resolution, reliability and availability of electron microscopy, including scanning (SEM) and transmittance (TEM) modes. The recent invention of scanning tunnelling (STM) and atomic force (AFM) microscopy, first appearing in 1982 and 1986, respectively,[1] also provided additional tools for probing and manipulating atomic scale objects.

SEM, AFM and STM instruments detect the surface features of nanostructured objects. However, when it comes to nanotubular morphology, sometimes the only method which can differentiate between the tubular or rod shape of objects is TEM imaging, which allows us to "see" the projection of the object. Furthermore, in order to avoid misinterpretation and artefacts associated with TEM imaging it is necessary to detect the nanotubes oriented parallel to the electron beam casting the projection in the shape of a circle. Although electron microscopy imaging allows us to *directly* detect the shape and dimensions of the objects with great accuracy, the disadvantage is that it covers only a limited area of the sample. As a result, there are always doubts as to how representative the imaged object is of the whole sample.

One of the methods which measures macroscopic parameters and associates them with microscopic parameters, is the method of gas adsorption into the porous samples. The most popular and developed method is nitrogen adsorption, which has been widely used in the investigation of catalysts for the characterisation of porous materials. In typical experiments, the adsorption of gaseous nitrogen on the surface of a material is studied at a temperature of – 196 °C, in the range of relative pressures from 0 to 1. Using an adsorption model, it is possible to determine the specific surface area (BET) and the pore size distribution (BJH). These data can then be associated with the morphology and size of the nanostructure. The advantage of the method is simplicity, cost and reproducibility. The disadvantage is that it requires a model which can associate pore size distribution with the dimensions of the nanostructure. This approach is usually applicable only to nanostructures from the same family (*e.g.* nanotubes, nanoparticles *etc*). Section 3.2 considers the porous structure of titanate nanotubes.

The degree of atomic order in nanostructures can be studied using electron or X-ray diffraction methods. There are some difficulties with the interpretation of diffraction data, due to the relatively small size (or polycrystallinity) of nanostructures resulting in a widening of the reflexes. Furthermore, some nanostructures (*e.g.* as-prepared TiO_2 anodized nanotubes) may be amorphous. In Section 3.1, crystallographic studies of titanate and TiO_2 nanotubes are reviewed.

The electronic structure and the nature of the surface atoms has been actively studied using conventional spectroscopy methods. Section 3.3 considers various spectroscopic data. Results from UV/VIS, PL, ESR, XPS, NMR, Raman and FTIR spectroscopy are interpreted.

The methods of computational chemistry are also widely used in nanoscience. In the instance of titanate nanotubes, however, their use is yet limited due to the massive number of atoms in its nanostructure, resulting in a huge basis set of atomic orbitals. For example, a typical 100 nm long titanate nanotube contains more than ten thousand titanium atoms, making direct *ab initio* calculations impossible using current computing facilities. However, future developments in computing technology and the implementation of various model approximations would allow further simulations and would stimulate progress in the discovery of new nanomaterials.

References

1. G. L Hornyak, H. F. Tibbals, J. Dutta and J. J. Moore, *Introduction to Nanoscience and Nanotechnology*, CRC Press, Taylor and Francis Group, 2009.
2. G. L. Hornyak, C. J. Patrissi, E. B. Oberhauser, C. R. Martin, J.-C. Valmalette, L. Lemaire, J. Dutta and H. Hofmann, *Nanostruct. Mater.*, 1997, **9**, 571.
3. M. Reibold, P. Paufer, A. A. Levin, W. Kochmann, N. Patzke and D. C. Meyer, *Nature*, 2006, **444**, 285.
4. C. Srinivasan, *Current Sci.*, 2007, **92**(3), 10.
5. R. J. Magyar, S. Tretiak, Y. Gao, H. L. Wang and A. P. Shreve, *Chem. Phys. Lett.*, 2005, **401**, 149.
6. L. A. Fendt, I. Bouamaied, S. Thoni, N. Amiot and E. Stulz, *J. Am. Chem. Soc.*, 2007, **129**, 15319.
7. L. Manna, E. C. Scher and A. P. Alivisatos, *J. Am. Chem. Soc.*, 2000, **122**, 12700.
8. A. P. Alivisatos, *J. Phys. Chem.*, 1996, **100**, 13226.
9. M. Monthioux and V. L. Kuznetsov, *Carbon*, 2006, **44**, 1621.

10. L. V. Radushkevich and V. M. Lukyanovich, *Russ. J. Phys. Chem. A*, 1952, **26**, 88.
11. S. Iijima, *Nature*, 1991, **354**, 56.
12. R. I. Jaffee and N. E. Promisel, (Ed.), *The science, technology, and application of titanium: Proceedings of the 1st International Conference on Titanium*, 1968, London, Pergamon Press, 1970.
13. A. Fujishima and K. Honda, *Nature*, 1972, **238**, 37.
14. A. Fujishima, K. Hashimoto and T. Watanabe, *TiO₂ Photocatalysis: Fundamentals and Applications*, BKC, USA, 1999.
15. T. Kasuga, M. Hiramatsu, A. Hoson, T. Sekino and K. Niihara, *Langmuir*, 1998, **14**, 3160.
16. X. Chen and S. S. Mao, *Chem. Rev.*, 2007, **107**, 2891.
17. D. V. Bavykin, J. M. Friedrich and F. C. Walsh, *Adv. Mater.*, 2006, **18**, 2807.
18. Q. Chen and L. -M. Peng, *Int. J. Nanotechnol.*, 2007, **4**, 44.
19. H. H. Ou and S. L. Lo, *Sep. Purif. Technol.*, 2007, **58**, 179.
20. C. A. Grimes, *J. Mater. Chem.*, 2007, **17**, 1451.
21. G. K. Mor, O. K. Varghese, M. Paulose, K. Shankar and C. A. Grimes, *Sol. Energy Mater. Sol. Cells*, 2006, **90**, 2011.
22. J. M. Macak, H. Tsuchiya, A. Ghicov, K. Yasuda, R. Hahn, S. Bauer and P. Schmuki, *Curr. Opin. Solid State Mater. Sci.*, 2007, **11**, 3.
23. C. Bae, H. Yoo, S. Kim, K. Lee, J. Kim, M. M. Sung and H. Shin, *Chem. Mater.*, 2008, **20**, 756.
24. Y. Ding and Z. L. Wang, *J. Phys. Chem. B*, 2004, **108**, 12280.
25. H. G. Yang and H. C. Zeng, *J. Am. Chem. Soc.*, 2005, **127**, 270.
26. D. V. Bavykin, V. N. Parmon, A. A. Lapkin and F. C. Walsh, *J. Mater. Chem.*, 2004, **14**, 3370.
27. Z. Y. Yuan, W. Zhou and B. L. Su, *Chem. Commun.*, 2002, **11**, 1202.
28. F. Keller, *J. Electrochem. Soc.*, 1953, **100**, 411.
29. H. Masuda and K. Fukuda, *Science*, 1995, **268**, 1466.
30. A. M. Azad, *Mater. Sci. Eng. A*, 2006, **435**, 468.
31. J. Zhang, S. Wei, J. Lin, J. Luo, S. Liu, H. Song, E. Elawad, X. Ding, J. Gao, S. Qi and C. Tang, *J. Phys. Chem. B*, 2006, **110**, 21680.
32. Y. Yang, X. Wang, C. Sun and L. Li, *Nanotech.*, 2009, **20**, 055709.
33. H. Deng, J. Wang, Q. Peng, X. Wang and Y. Li, *Chem. -Eur. J.*, 2005, **11**, 6519.
34. L. Zhuo, J. Ge, L. Cao and B. Tang, *Cryst. Growth Design*, 2009, **9**, 1.
35. L. Z. Pei, H. S. Zhao, W. Tan and Q. F. Zhang, *J. Appl. Phys.*, 2009, **105**, 054313.
36. H. Tsuchiya and P. Schmuki, *Electrochem. Commun.*, 2005, **7**, 49.
37. C. J. Jia, L. D. Sun, Z. G. Yan, L. P. You, F. Luo, X. D. Han, Y. C. Pang, Z. Zhang and C. H. Yan, *Angew. Chem. Int. Ed.*, 2005, **44**, 4328.
38. Y. Yang, X. Wang, C. Zhong, C. Sun and L. Li, *Appl. Phys. Lett.*, 2008, **92**, 122907.
39. W. Fan, S. Sun, L. You, G. Cao, X. Song, W. Zhang and H. Yua, *J. Mater. Chem.*, 2003, **13**, 3062.

40. D. Zheng, S. Sun, W. Fan, H. Yu, C. Fan, G. Cao, Z. Yin and X. Song, *J. Phys. Chem. B*, 2005, **109**, 15439.
41. X. Wang and Y. Li, *Chem. Lett.*, 2004, **33**, 48.
42. R. Ma, Y. Bando and T. Sasaki, *J. Phys. Chem. B*, 2004, **108**, 2115.
43. K. Zhang, J. Wang, X. Lu, L. Li, Y. Tang and Z. Jia, *J. Phys. Chem. C*, 2009, **113**, 142.
44. G. B. Saupe, C. C. Waraksa, H. N. Kim, Y. J. Han, D. M. Kaschak, D. M. Skinner and T. E. Mallouk, *Chem. Mater.*, 2000, **12**, 1556.
45. M. E. Spahr, P. Bitterli, R. Nesper, M. Muller, F. Krumeich and H. U. Nissen, *Angew. Chem. Int. Ed.*, 1998, **37**, 1263.
46. Y. Wang and G. Cao, *Chem. Mater.*, 2006, **18**, 2787.
47. M. E. Spahr, P. Stoschitzki-Bitterli, R. Nesper, O. Haas and P. Novak, *J. Electrochem. Soc.*, 1999, **146**, 2780.
48. M. Niederberger, H. J. Muhr, F. Krumeich, F. Bieri, D. Gunther and R. Nesper, *Chem. Mater.*, 2000, **12**, 1995.
49. J. Zhang, L. Sun, C. Liao and C. Yan, *Chem. Commun.*, 2002, 262.
50. H. Tsuchiya, J. M. Macak, A. Ghicov, L. Taveira and P. Schmuki, *Corros. Sci.*, 2005, **47**, 3324.
51. C. N. R. Rao and M. Nath, *Dalton Trans.*, 2003, **1**, 1.
52. H. Cui, H. Liu, X. Li, J. Wang, F. Han, X. Zhang and R. I. Boughton, *J. Solid State Chem.*, 2004, **177**, 4001.
53. P. Afanasiev, C. Geantet, C. Thomazeau and B. Jouget, *Chem. Commun.*, 2000, 1001.
54. Y. Tian, Y. He and Y. Zhu, *Mater. Chem. Phys.*, 2004, **87**, 87.
55. H. A. Therese, J. Li, U. Kolb and W. Tremel, *Solid State Sci.*, 2005, **7**, 67.
56. J. Chen, Z. L. Tao and S. L. Li, *Angew. Chem. Int. Ed.*, 2003, **42**, 2147.
57. N. Roveri, G. Falini, E. Foresti, G. Fracasso, I. G. Lesci and P. Sabatino, *J. Mater. Res.*, 2006, **21**, 2711.
58. B. Jancar and D. Suvorov, *Nanotechnology*, 2006, **17**, 25.
59. L. Ren, J. S. Hu, L. J. Wan and C. L. Bai, *Mater. Res. Bull.*, 2007, **42**, 571.
60. H. Yang, C. Wang and Z. Su, *Chem. Mater.*, 2008, **20**, 4484.
61. A. McDonald, B. Scott and G. Villemure, *Microporous Mesoporous Mater.*, 2009, **120**, 263.
62. E. N. Korytkova, L. N. Pivovarova, I. A. Drosdova and V. V. Gusarov, *Russ. J. Gen. Chem.*, 2007, **77**, 1669.
63. X. Wang, J. Zhuang, J. Chen, K. Zhou and Y. Li, *Angew. Chem. Int. Ed.*, 2004, **43**, 2017.
64. W. Dong, W. Li, K. Yu, K. Krishna, L. Song, X. Wang, Z. Wang, M. O. Coppens and S. Feng, *Chem. Commun.*, 2003, 1302.
65. Y. Li, J. Wang, Z. Deng, Y. Wu, X. Sun, D. Yu and P. Yang, *J. Am. Chem. Soc.*, 2001, **123**, 9904.
66. C. N. R. Rao and A. Govindaraj, *Nanotubes and Nanowires*, Royal Society of Chemisty, 2005.

CHAPTER 2

Synthesis Techniques and the Mechanism of Growth

The methods for nanostructured TiO_2 synthesis can be split into two categories: templated and non-templated procedures (see Figure 2.1). In the case of elongated TiO_2 nanostructures, there are two common non-templated methods of synthesis, namely alkaline hydrothermal synthesis and electrochemical anodising of titanium. These methods are reviewed below.

2.1 Template Methods

The method of template synthesis of nanostructured materials, being a classical "bottom-up" method, has become extremely popular during the last decade.[1] The method utilises the morphological properties of known and characterised materials (templates) in order to construct materials having a similar morphology by methods including reactive deposition. This method is very general; by adjusting the morphology of template material, it is possible to prepare numerous new materials having a regular, uniform and controlled morphology on the nano- and micro scale. A disadvantage is that, in most cases, the template material is sacrificial and needs to be destroyed after synthesis leading to increased cost of materials; as in the case of all surface finishing techniques, it is also important to maintain a high level of surface cleanliness to ensure good adhesion between the substrate and the surface coating.

The method can be split in several stages (see Figure 2.2). The first stage is the deposition of required materials onto the surface and into the pores of the templated substrate. There are several approaches for deposition of nanostructured material precursors including: precipitation of sol–gel synthesised materials from solutions onto the surface of the template; atomic layer deposition (ALD) from the gas or liquid phase; chemical vapour deposition (CVD) of required materials from the gas phase; electrodeposition, photodeposition and thermosetting stimulated by electrical current, absorbed light

RSC Nanoscience & Nanotechnology No. 12
Titanate and Titania Nanotubes: Synthesis, Properties and Applications
By Dmitry V. Bavykin and Frank C. Walsh
© Dmitry V. Bavykin and Frank C. Walsh 2010
Published by the Royal Society of Chemistry, www.rsc.org

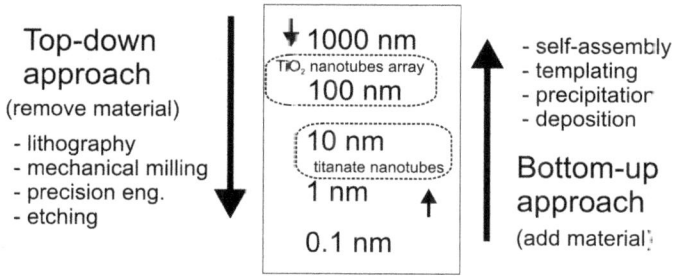

Figure 2.1   Top-down and bottom-up approaches for the preparation of nanostructured materials.

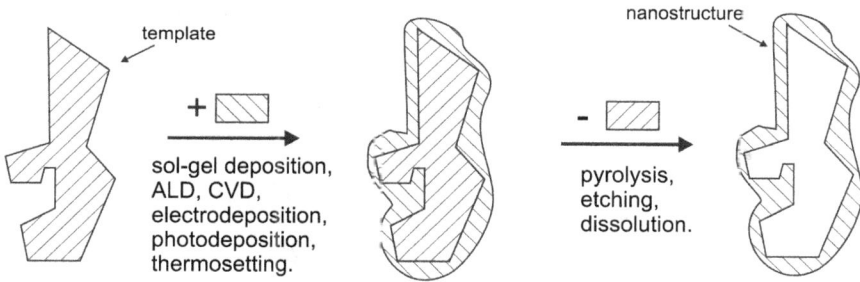

Figure 2.2   The template method for the preparation of nanostructured materials. (ALD: atomic layer deposition; CVD: chemical vapour deposition).

and temperature, respectively. After successful adhesion of material, the template is removed by various methods, including pyrolysis, selective etching or dissolution.

The templates can be classified into several types according to the geometry of the deposited layer.[2] In instances where the precursor is deposited onto the external surface of the template (*e.g.* the surface of nanofibres, nanorods or spheres), the template is classified as positive. In instances where the precursor is deposited into the internal pores of the template, (*e.g.* anodic aluminium oxide – an AAO membrane), the template can be classified as negative. The types of template can be also divided according to the characteristic size of the nanostructure, allowing templates to be categorised into macroporous, mesoporous and microporous types.[3]

An example of a macroporous negative template is provided by polystyrene beads with a narrow distribution of diameter, allowing their self assembling into photonic structures with regular pores having a characteristic pore size in the range of several tens of nanometres. Sol–gel deposition of TiO$_2$ into these pores, followed by removal of resin under calcination, results in the formation of an inversed opal[4] type nanostructured TiO$_2$ (see Figure 2.3).

Template   TiO₂ nanostructure

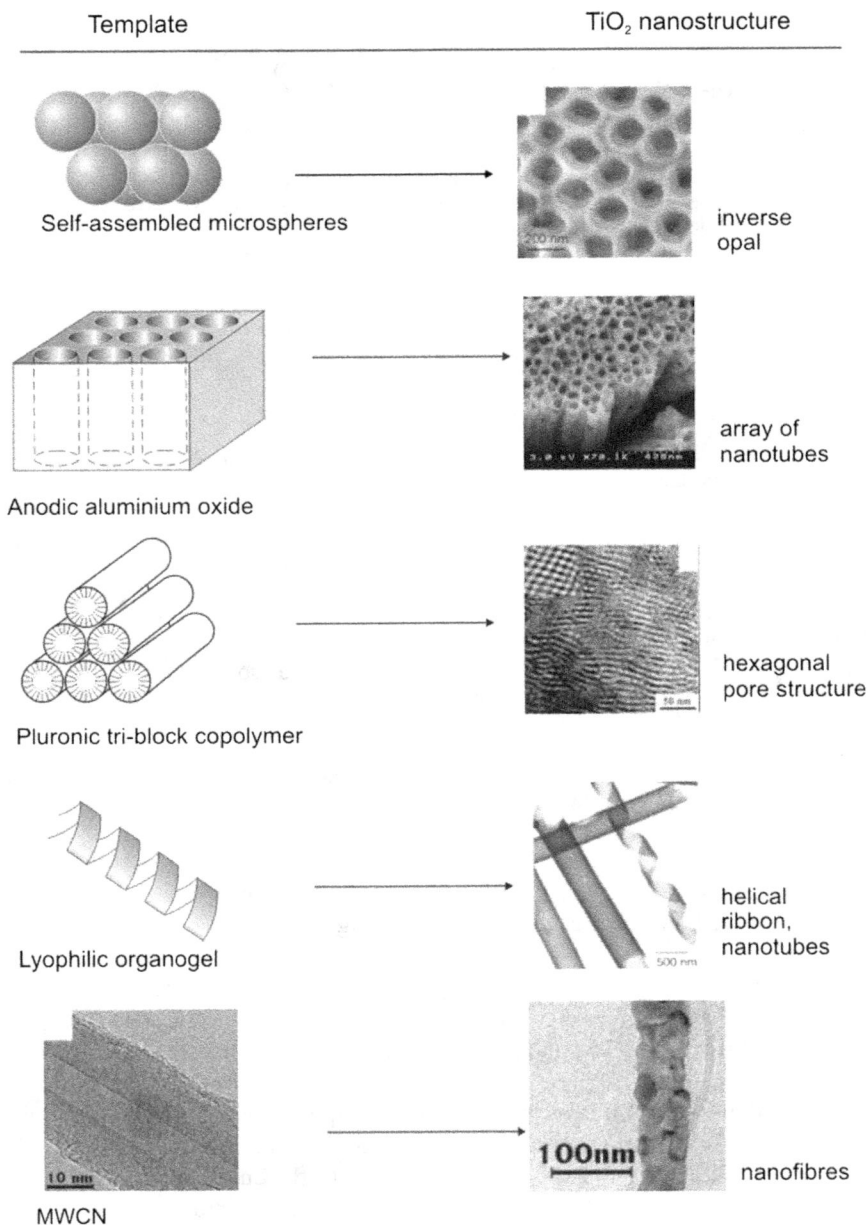

Self-assembled microspheres   inverse
                                                            opal

Anodic aluminium oxide   array of
                                                            nanotubes

Pluronic tri-block copolymer   hexagonal
                                                            pore structure

Lyophilic organogel   helical
                                                            ribbon,
                                                            nanotubes

MWCN   nanofibres

Figure 2.3   Important examples of templates used to prepare nanostructured TiO₂
             materials. (Images are reproduced with kind permission as follows:
             inversed opals from ref 4. nanotubular array from ref 16. tri-block
             copolymer template TiO₂ from ref 8. helical ribbon TiO₂ from ref. 5 and
             MWCN-templated TiO₂ nanofibres from ref. 6).

An example of a macroporous positive template is the lyophylic organogel which can form tubular or helical nanostructures over a certain concentration range.[5] The controlled hydrolysis of titanium(IV) esters in the presence of an organogel may result in the precipitation of TiO_2 on the surface of the template and a reproduction of its morphology. After removal of the organic phase, the TiO_2 produced has a nanotubular or helical morphology (see Figure 2.3).

Multi-walled carbon nanotubes (MWCN)[6] or carbonaceous nanofibres[7] can be considered as mesoporous positive templates. Sol–gel deposition of TiO_2 on the external surface of MWCN or nanofibres followed by calcination in air results in the formation of TiO_2 nanofibres or nanotubes (see Figure 2.3 and Table 2.1).

A new class of triblock copolymer with the general formula $HO(CH_2$-$CH_2O)_n(CH_2CH(CH_3)O)_m(CH_2CH_2O)_nH$ has been recently introduced. In particular, solutions above their critical micelle concentration can give rise to self-assembled, hexagonal, packed-rod like structures (see Figure 2.3). The sol–gel hydrolysis/condensation of titanium(IV) alkoxides in the presence of such triblock copolymers [*e.g.* Pluronic P123 with $n = 20$, $m = 70$ (ref. 8) or Pluronic F127 with $n = 106$, $m = 70$ (ref. 9)], followed by polymer removal at elevated temperatures, results in the formation of mesoporous TiO_2 with a hexagonal pore structure.

The preparation of TiO_2 nanotubes by chemical templating[2] usually involves the controlled sol–gel hydrolysis of solutions of titanium containing compounds in the presence of templating agents, followed by polymerisation of TiO_2 in the self-assembled template molecules or deposition of TiO_2 onto the surface of template aggregates. The next stages are the selective removal of the templating agent and calcination of the sample.

The group consisting of self-assembled organic surfactant template molecules is probably the largest group, since a wide range of organic molecules can be used. Among the organic templates used for TiO_2 nanotube synthesis, are the organogel of *trans*-(1R,2R)-1,2-cyclohexanedi(11-aminocarbonylunde-cylpyridinium)[5] and dibenzo-30-crown-10-appended cholesterol[10] and the hydrogel of tripodal cholamide-based materials,[11] together with laurylamine hydrochloride surfactant.[12]

Other examples of specialised templating agents include tobacco mosaic viruses,[13] precipitated platinum salts[14] and electrospun polyacrylonitrile fibres.[15]

Porous alumina, produced by anodising of aluminium foil (anodic aluminium oxide AAO), has been widely used as a macroporous negative template for the preparation of TiO_2 nanotubes. The internal surface of cylindrical pores of AAO is used for the deposition of TiO_2 thin films from various precursors by sol–gel depositions[16–18] or electrodeposition.[19] After selective removal of alumina using a concentrated solution of NaOH, the external diameter of TiO_2 hollow fibres corresponds to the diameter of alumina pores. The internal diameter of the TiO_2 nanotubes depends on the synthesis conditions and the thickness of the wall (Table 2.1).

After calcination at 500 °C, the crystal structure of TiO_2 nanotubes produced by templating is usually amorphous or polycrystalline anatase. The tubes have

Table 2.1 Morphological properties of TiO$_2$ nanotubes produced by the sol–gel method in the presence of templating agents.

Precursor	Template	Conditions	Nanotube diameter/nm	Ref.
Ti(O*i*Pr)$_4$		25 °C, ethanol, NH$_4$OH	50–300	5
Ti(O*i*Pr)$_4$		25 °C, 1-butanol, benzylamine	500	10
Ti(OBu)$_4$		25 °C, ethanol, CH$_3$COOH, H$_2$O	4–7	11
Ti(OBu)$_4$	CH$_3$(CH$_2$)$_{11}$NH$_2$ · HCl	25–40 °C, H$_2$O	1800–6000	12
Ti(O*i*Pr)$_4$	Tobacco mosaic viruses	25 °C, ethanol	20	13
Ti(OBu)$_4$	[Pt(NH$_3$)$_4$](HCO$_3$)	25 °C, ethanol	100	14
TiCl$_4$	polyacrylonitrile fibres	25 °C, ethanol, NH$_4$OH	220	15
Ti(OBu)$_4$	Carbonaceous nanofibres	25 °C, ethanol	20	17
Ti(O*i*Pr)$_4$	AAO membrane	25 °C, pressure impregnation	60–70	16
TiF$_4$	AAO membrane	60 °C, HCl	2.5–5, 70–100	18
Ti(O*i*Pr)$_4$	AAO membrane	25 °C, ethanol, CH$_3$COOH	120–140	17
TiCl$_3$	AAO membrane	25 °C, HCl, electrodeposition	70–100	19

AAO = Anodic aluminium oxide, Bu = *n*-butyl and *i*Pr = isopropyl.

different mean internal diameters depending on the nature of the templating agent (Table 2.1). The specific surface area for very wide tubes is not very high except for the cases where the structure of the microtube wall contains small channels.[12] Nanotubes have potential uses in the photocatalytic removal of organic pollutants or in solar cells,[20,21] although industrial applications may be limited by the cost of materials, insufficient material characterisation and concerns over long term instability.

2.2 Alkaline Hydrothermal Synthesis of Elongated Titanates

Titanium dioxide can be considered as an amphoteric oxide, since it can either react with strong acid forming titanium(IV) salts (*e.g.* titanium(IV) chloride, $TiCl_4$; or titanil sulfate, $TiOSO_4$), or with strong bases forming titanate salts with the general formula $M_{2n}Ti_mC_{2m+n} \cdot xH_2O$ (where M is an alkaline metal cation and n, m and x are integers; *e.g.* $Na_2Ti_6O_{13}$). Whilst the solubility of TiO_2 in strong acid can be relatively high, it is usually much lower in basic solutions. However, the solubility can be increased by increasing the temperature of solution. Traditionally, titanates are produced by either solid state or melt reaction between TiO_2 and a second metal oxide or carbonate at elevated temperatures (> 1000 °C), to form bulk titanate crystals. By contrast, the alkaline hydrothermal treatment of TiO_2 at elevated temperatures and pressure can provide an alternative route for the synthesis of nanostructured titanates, *via* a sequence involving the dissolution of the initial TiO_2 and the crystallisation of the final product.

2.2.1 Alkaline Hydrothermal Synthesis of Titanate Nanotubes and Nanofibres

In 1998, Kasuga *et al.*[22] first reported a simple method for the preparation of TiO_2 nanotubes without the use of sacrificial templates, involving the treatment of amorphous TiO_2 with a concentrated solution of NaOH (10 mol dm^{-3}) in a PTFA-lined batch reactor at elevated temperatures. In a typical process, several grams of TiO_2 raw material can be converted (< 100%) to nanotubes at temperatures in the range of 110–150 °C, followed by washing with water and HCl (0.1 mol dm^{-3}). It has since been demonstrated that all polymorphs of TiO_2 (anatase, rutile,[23,24] brookite[25] or amorphous forms) can be transformed to nanotubular or nanofibrous TiO_2 under alkaline hydrothermal conditions.

Figure 2.4 shows typical TEM images of titanate nanotubes and nanofibres obtained by such treatments. Generally, the nanotubes have a multi-wall morphology (with typically four walls) and a distance between successive layers of approximately 0.72 nm (in the protonated form). The inner diameter of nanotubes is in range from 2 to 20 nm and the majority of tubes are open at both ends. Each tube tends to have a constant diameter along its length. The distribution of internal diameters in the sample of nanotubes is relatively wide

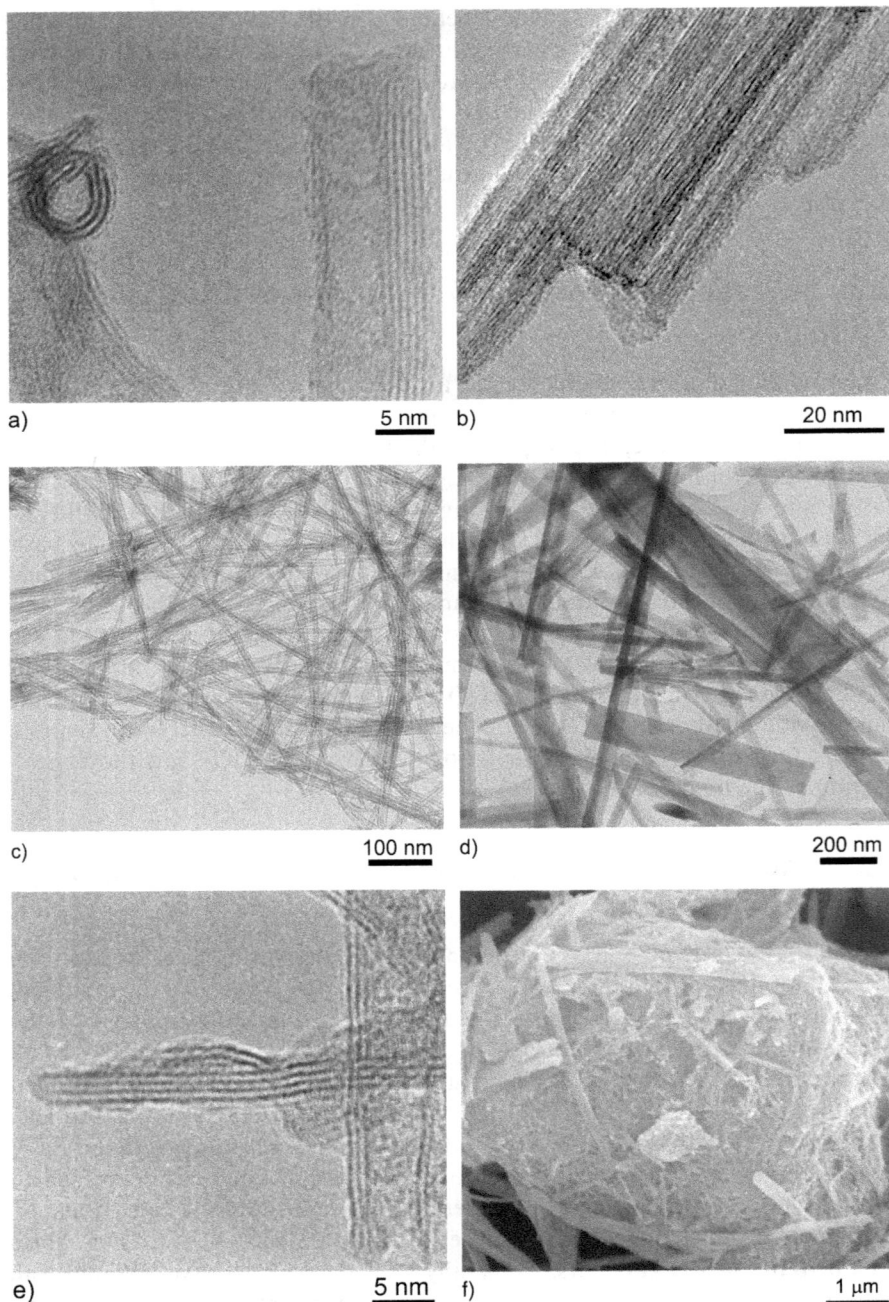

Figure 2.4   TEM images of a) and c) titanate nanotubes, b) and d) nanofibres and e) multilayer nanosheets, produced by the alkaline hydrothermal treatment of anatase with NaOH ($10\,mol\,dm^{-3}$) at $140\,°C$ (for nanotubes and nanosheets) and at $190\,°C$ (for nanofibres). A SEM image of f) an agglomerate of titanate nanotubes. (Images for a), c), d), e) and f) are reproduced with kind permission from ref. 29).

when compared with those of the nanotubes produced using templating methods. Figure 2.4a shows the radial cross-section of a selected nanotube (left hand side) having a pronounced seam, and the axial cross-section of another nanotube (right hand side) having an asymmetrical number of walls. The nanotubes can be long – up to several microns, resulting in aspect ratios of several order of magnitude. Nanotubes are randomly assembled into agglomerate particles as seen in Figure 2.4f. Their size and shape depends on the synthetic conditions. The typical size of agglomerates is several microns. Nanotubes can be isolated from the bundles into an aqueous colloidal solution using surfactants and mild physical treatment (*e.g.*, by stirring or ultrasound).

Nanofibres, produced by hydrothermal treatment of TiO_2 with NaOH ($10\,mol\,dm^{-3}$) at temperatures higher than 170 °C, are characterised by a solid elongated morphology with a typical width being in the range of 20 to 200 nm and a length exceeding several microns, as shown in Figures 2.4b and 2.4d. These samples of nanofibres are also characterised by a wide distribution of nanofibre widths and lengths. The nanofibres also have a layered structure with a characteristic interlayer distance of approximately 0.72 nm (in the protonated form).

Sometimes, it is possible to observe partially wrapped multi-layer nanosheets, however, each sample of nanotubes contains a small amount of non-wrapped multilayered nanosheets, as shown in Figure 2.4e. It is believed that these nanosheets play an important role in the mechanism of nanotube formation.

2.2.2 Mechanism of Nanostructure Growth

The key to developing and exploiting new nanostructured materials lies in an increased understanding of how synthesis conditions affect the properties of nanostructured materials in order to tailor materials to specific needs. A knowledge of the mechanism of nanostructure formation is of particular importance.

Nanotubes

Since the discovery of the wet chemical method of transformation of raw titania to nanotubular titanates, many attempts have been made to describe this mechanism and to rationalise these transformations. Current ideas concerning the mechanism for the formation of nanotubular titanates are reviewed below. Originally, Kasuga *et al.*[22] considered that nanotubular morphology occurred during post hydrothermal acid washing. Some researchers still support this suggestion,[26] but it was clearly demonstrated, by washing samples with ethanol or acetone,[27,28] that nanotubular sodium titanates are formed *during* alkaline hydrothermal processing.

Unfortunately, it has not proved possible to adapt the concept of the catalytic growth of hollow fibres on the metal nanoparticles of catalysts, where the diameter of fibres corresponds to the diameter of the particles involved in the preparation of carbon nanotubes. Indeed, during the alkaline hydrothermal synthesis of titanate nanotubes, there is no possible candidate to play the role of a size-determining nanoparticle. Unlike carbon nanotubes, which are often produced by the catalytic pyrolysis of hydrocarbons, titania nanotubes produced *via* the alkaline hydrothermal method have never been observed in a single layer form. All reports of TiO_2 nanotubes describe these samples as nanotubular structures with multilayered walls.

Although the detailed sequence of events is still under discussion, it is clear that, during transformation of TiO_2 (anatase, rutile or amorphous) under alkaline conditions, the observed intermediate single-layer and multilayered titanate nanosheets play a key role in the formation of tubular morphology.[29,30] These nanosheets can scroll or fold into a nanotubular morphology. The driving force for curving these nanotubes has been considered by several groups.

Zhang *et al.*[31] considered that single surface layers experienced an asymmetrical chemical environment, due to the imbalance of H^+ or Na^+ ion concentrations on the two different sides of a nanosheet, giving rise to an excess surface energy, resulting in bending. The system could be presented as a plane with two springs on each side parallel to this plane (crystallographic axes c and b; Figure 2.5). When both sides have a symmetrical chemical environment, both spring constants have similar values. As a result, all tensions are compensated and the plane is straight. When the trititanate nanosheet has an asymmetric

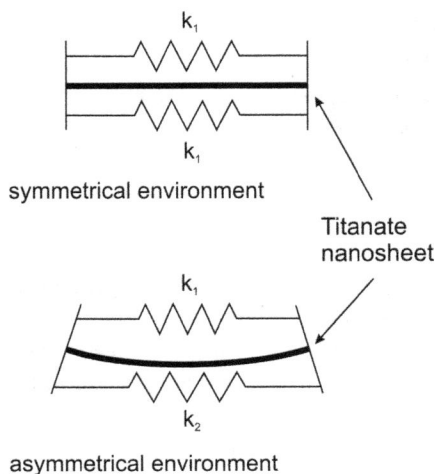

Figure 2.5   The driving force for bending titanate nanosheets under alkaline hydrothermal conditions. An asymmetrical chemical environment results in a
difference in surface tension on each side of the nanosheet; k_1 and k_2 are
the spring constants for each side.

proton distribution, then each side has a different free surface energy value (spring constant) and, in order to compensate for the imbalance in the surface tensions, the plane bends towards the surface with the higher spring constant value. During the bending process, a strain energy arises which prevents work against bending. In a simple approach, the excess of energy of the layer (E_{layer}) can be expressed as the difference between two terms:

$$E_{layer} = \frac{\alpha}{r^2} - \frac{\beta}{r} \qquad (2.1)$$

where r is the radius of curvature of the curved nanosheet, α and β are the proportionality constants responsible for the elasticity of the nanosheet and the imbalance in surface tension, respectively. The detailed analysis of the structural changes of nanosheets and *ab initio* DFT calculations of the total energy of a curved fragment of trititanate nanosheets as a function of curvature radius,[32] demonstrates that the optimal radius of nanotubes is *ca.* 4 nm and the number of layers in the final titanate nanotube is four. However, these results contradict the recently observed dependence of the average nanotube diameter on conditions of synthesis *e.g.* temperature or the ratio of TiO_2 to NaOH.[26,33]

Another reason for the bending of multilayered nanosheets,[29] is that mechanical tensions arise during the process of dissolution–crystallisation in nanosheets.[34] In a layered chemical compound, the interaction energy between atoms in neighbouring layers is usually less than that between atoms in the same layer. The growth of nanosheets predominately occurs at the edges of the individual layers, rather than in the initiation of a new layer. During spontaneous crystallisation and the rapid growth of layers, it is possible that the width of the different layers varies as shown in Figure 2.6a. It is likely that the imbalance in the layer width (Δx) creates a tendency of layers to move within the multi-walled nanosheet in order to decrease the excess surface energy (see Figure 2.5b). This can result in the bending of multilayered nanosheets, as seen in Figure 2.6c. It was demonstrated by simplified considerations that, during the simultaneous shift of layers and the bending of a nanosheet, the gain in surface energy can be sufficient to compensate for the mechanical tensions arising in the material during curving and wrapping into nanotubes. This statement can be described by the inequality:

$$\frac{k(\Delta y)^2}{2} < \sigma \cdot L \cdot \Delta x \qquad (2.2)$$

where the left side corresponds to the mechanical energy of the bent nanosheet, and the right side corresponds to the excess surface area associated with an imbalanced layer, Δy is the curving deformation of the layer, k is the spring constant, σ is the excess surface energy per unit area at the solution–particle interface and L is the length of the imbalanced layer in the nanosheet. A detailed analysis[29] of the inequality (2.2) showed that a change in surface

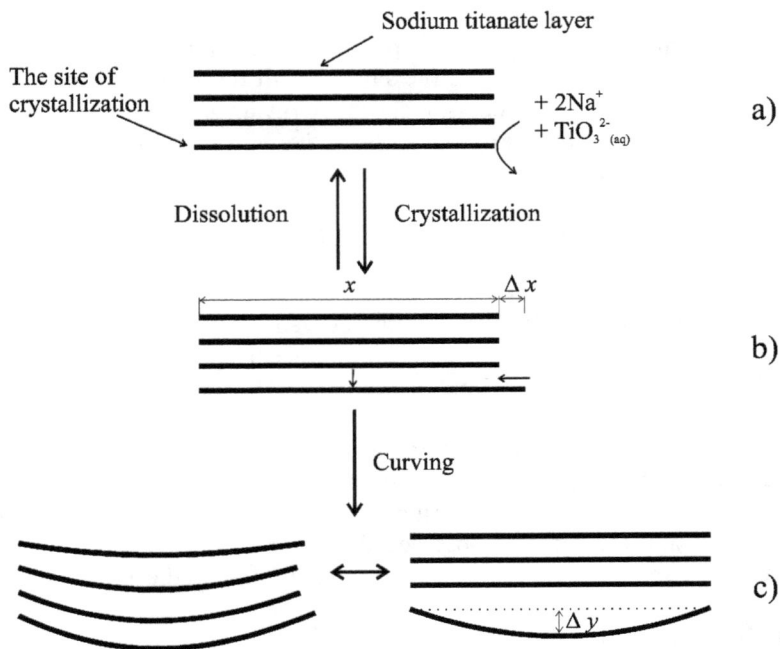

Figure 2.6 The driving forces for bending multilayered titanate nanosheets under alkaline hydrothermal conditions due to an imbalance in layer widths, Δx, resulting in shifting of layers and bending of nanosheets.

energy for one nanosheet is approximately 500×10^{-18} J, whereas the change in elastic energy is approximately 1×10^{-18} J. Thus, the gain in surface energy is sufficient to compensate for the mechanical tensions arising in the material during curving and wrapping into nanotubes.

In the absence of factors which stabilise the bent form of nanosheets, the reverse process of transformation of bent nanosheets back to the planar form should occur. While it is difficult to find the stabilisation factors for a curved single-layer nanosheet, (and unless an asymmetrical environment is maintained) the curved form of multilayered nanosheets could be stabilised by periodic potentials in the crystal lattice.

In order to visualise such stabilization, let us consider a model of nanosheets consisting of only two layers with n and $n+m$ elements in each layers as seen in Figure 2.7. Each element (shown as a ball) represents the motifs of the crystalline titanates, which could be TiO_6 octahedra. a and b are the distances between layers and elements within one layer, respectively, which can be associated with the lattice parameters of titanates (see Chapter 3).

Assuming that during bending of multilayered nanosheets the distances between layers and between motifs within one layer are constant, one can describe such bending as a gradual decrease of the radius of curvature of the nanosheets from infinity (flat nanosheet) to a certain value r. The coordinates of the ball can be expressed as:

$$\begin{pmatrix} x1_i \\ y1_i \end{pmatrix} = \begin{pmatrix} r \times \cos(\frac{b*i}{r}) \\ r * \sin(\frac{b*i}{r}) \end{pmatrix}, \; i = 0, \; n-1 \tag{2.3}$$

and

$$\begin{pmatrix} x2_j \\ y2_j \end{pmatrix} = \begin{pmatrix} (r+a) * \cos(\frac{b*j}{r+a}) \\ (r+a) \times \sin(\frac{b*j}{r+a}) \end{pmatrix}, \; j = 0, \; n+m-1 \tag{2.4}$$

where $x1_j$ and $y1_i$ are the coordinates of i balls in the first layer and $x2_j$ and $y2_j$ are the coordinates of j balls in the second layer. In layered compounds, the interaction energy between layers is usually much weaker than the interaction energy between atoms in the layer. It is possible to describe such weak interaction as Van der Waals type using Lennard–Jones potentials. The interaction energy between all balls in Figure 2.7 can then be expressed as:

$$W = \varepsilon \sum_{i=0}^{n-1} \sum_{j=0}^{n+m-1} \left[\left(\frac{a}{\sqrt{(x1_i - x2_j)^2 + (y1_i - y2_j)^2}} \right)^{12} - 2 \left(\frac{a}{\sqrt{(x1_i - x2_j)^2 + (y1_i - y2_j)^2}} \right)^{6} \right] \tag{2.5}$$

where ε is the coefficient characterising the strength of the interactions. Figure 2.8 shows the total interaction energy, W, between two layers in the nanosheets as a function of the radius of curvature of the nanosheet. The value of W was calculated by inserting Equations (2.3) and (2.4) into Equation (2.5), followed by summing over all i and j balls. The parameters a and b were selected to be as close as possible to the lattice parameters of layered titanates.

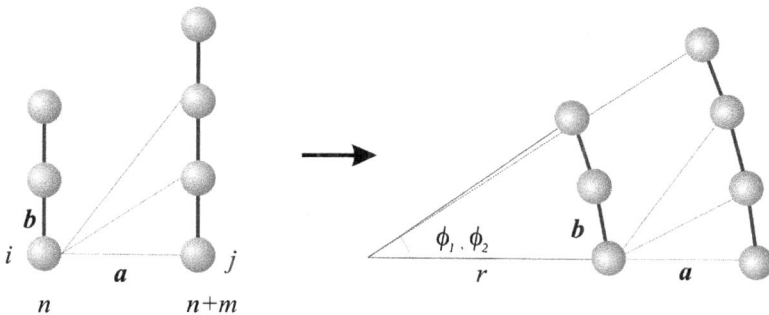

Figure 2.7 A model for bending of an imbalanced two-layer nanosheet with n and $n+m$ elements in each layer; a and b are lattice parameters, r is the radius of curvature and ϕ_1 and ϕ_2 are arc angles of the first and second layers, respectively.

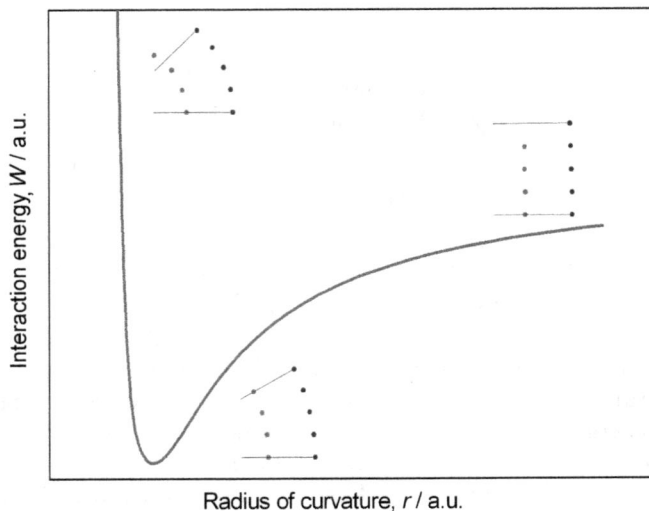

Figure 2.8   Interaction energy between the elements in two nanosheet layers (with n and $n+m$ elements in each layer) as a function of the radius of curvature of the nanosheet. The interaction energy between two elements from different layers is described by the Lennard–Jones potential.

The results of the calculations show that the bending of two imbalanced layers in the nanosheet results in a decrease of the total energy. This corresponds to the decrease of r from infinity towards a smaller curvature radius (see Figure 2.8). When the radius of curvature becomes too small, the additional bending will result in a sharp rise in energy, pushing the system into an energetically unfavourable state. It is important to note that at the point of minimum energy, the arc angles for the first and second layers are approximately equal to each other, meaning that the edge balls are situated in radial lines.

The radius of minimum interaction energy depends on the parameters n and m. These parameters can be associated with the kinetic rate of nanosheet crystallization, which can be controlled by the conditions of nanotube growth. This is in agreement with the experimental observation that the kinetic rate of curving of nanotubes might control the diameter of the resultant nanotubes.[29]

In contrast to the asymmetrical chemical environment, the mechanism of multilayer nanosheets bending cannot be applied to single-layer nanosheets, which remain unfolded in aqueous solution according to experimental observations.[29,35] The bending of multilayer nanosheets[36] may, however, result in the formation of several different types of nanotubes, depending on the method of loop closing (see Figure 2.9). The ideal sealing of matching layers ends in scrolled multilayer nanosheets would result in the formation of a seamless cross-section structure composed of concentric rings (as shown in Figure 2.9b for chrysotile nanotubes).[37] The sealing of non-matching layers results in the possibility of a nanosheet forming "snail" type scrolls, which can also be

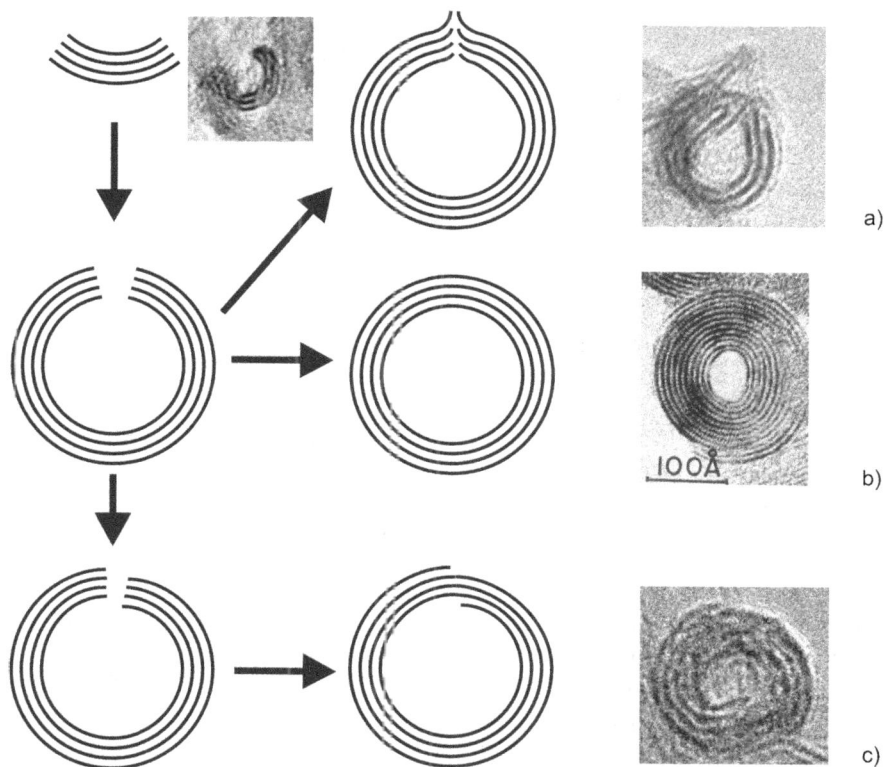

Figure 2.9 Three methods of loop closing resulting in a) "onion," b) "concentric" and c) "snail" type nanotubes. (Images are reproduced with kind permission as follows: a) from ref. 29 b) from ref. 39 and the bent multilayer nanosheet in the top left corner from ref. 36).

formed by the helical scrolling of a single-layer nanosheet (as seen in Figure 2.9c). Sometimes, the sealing of nanosheet ends cannot be completed due to steric restraints, resulting in the formation of "onion"-type structures in which the continuous seam is seen along the length of the nanotube.

It is believed that the existence of nanosheets is crucial to the subsequent formation of nanotubes. It is remarkable that, starting from colloidal single-layered isolated trititanate nanosheets, titanate nanotubes can be readily produced in alkaline conditions at room temperatures.[30] In contrast, starting from bulk-layered sodium trititanate ($Na_2Ti_3O_7$) in hydrothermal alkaline conditions, the formation of nanotubes is not observed.[38] However, after several days of hydrothermal treatment (without further additions of NaOH) at temperatures in the range 140–170 °C the resultant titanate nanotubes are characterised by very wide diameter (several tens of nanometres).[39] Such results suggest that not only is the presence of layered titanates in the reaction mixture important, but so too is their morphology.

Nanofibres

Under alkaline hydrothermal conditions, nanosheets can also be converted to nanofibres instead of scrolling into nanotubes. This usually occurs at temperatures above 170 °C or when KOH is used in place of NaOH. In both cases, the concentration of dissolved Ti(IV) was found to be similar to, but higher than, that in the case of nanotube synthesis.[58,62] An increase in the local concentration of Ti(IV) may result in a faster rate of nanosheet growth, with less effect on the rate of nanosheet scrolling. In this case, if the rate of crystallisation is large enough, the thickness of nanosheets can exceed a particular value where they become too rigid to bend before curving can occur. This will result in the formation of nanofibres rather than nanotubes.

Energetically, nanofibres are thermodynamically more stable than nanotubes since the latter have an increased surface area and higher stresses within the crystal lattice. It has been reported that long-term alkaline hydrothermal treatment of TiO_2[40] or over intensification of the synthesis conditions by the use of a revolving autoclave,[51] the addition of nanotube seeds[41] or microwave radiation[42] may result in favourable conditions for the formation of nanotubes, but nanofibres may be formed instead. This suggests that kinetic factors can exert a strong control over the route of nanosheet transformation.

It is interesting to note that the axis of nanotubes (direction [010]) does not always coincide with the axis of nanofibres (direction [001]). This provides an insight into the mechanism of titanate nanotube growth. The fact that nanofibres prefer to crystallise along the crystallographic axis c[43] suggests that the rate of dissolution–crystallisation along this axis is maximised. Under certain conditions, an imbalance in nanosheet width is expected along the c axis. Thus, a bending of nanosheets will occur around axis b. When the curved nanosheet closes the loop (a nanoloop[38] or rather short tubes), the direction of fastest growth disappears. There will be only two directions for nanotube growth: the radial direction (along crystallographic axis a) and the axial direction (along crystallographic axis b). Kukovecz *et al.*[38] propose that the nanoloops provide seeds for the further growth of nanotubes along the preferred direction of growth of axis b. If the nanosheet rolls up into conical shaped tubes, then further growth will result in the formation of closed-end nanotubes. At higher temperatures, bending of nanosheets does not occur readily and the resulting nanofibres are long in the crystallographic direction c.

Overall, the process of transformation of raw TiO_2 to nanotubular titanate can be considered to take place in several stages: (1) the slow dissolution of raw TiO_2, accompanied by the epitaxial growth of layered nanosheets of sodium trititanates; (2) the exfoliation of nanosheets; (3) the crystallisation of dissolved titanates on nanosheets, resulting in mechanical tensions which induce the curving and wrapping of nanosheets to nanotubes; (4) the growth of nanotubes along the length; and (5) the exchange of sodium ions by protons during washing and the separation of nanotubes. The sequence of events during titanate nanofibre growth is similar except for the absence of the curving step.

2.2.3 Methods to Control the Morphology of Nanostructures

Since the introduction of the alkaline hydrothermal synthesis of titanate nanotubes,[22] many efforts have been made to adapt this technique to more suitable technological processes allowing an easy and low-cost scale up of production. Potential routes for the preparation of titanate nanotubes are shown in Figure 2.10. The original method (route 1 in Figure 2.10) includes the use of TiO_2 raw material in aqueous NaOH ($10 \, mol \, dm^{-3}$) at temperatures in the range of 110 to 150 °C for 24 hours. The form of the TiO_2 reactant can include anatase, rutile, amorphous TiO_2, or even Ti metal. The choice of initial raw material may affect the morphology of the resultant nanotubes, but no systematic data on this subject is available. The hydrothermal method traditionally requires the use of an autoclave with chemically resistant vessels (usually lined with PTFE), in order to withstand such a concentrated and hot alkaline environment. The advantages of this method, however, are that it involves a single-stage process and relatively low hydrothermal temperatures are required to achieve an essentially complete conversion of initial raw materials into titanate nanotubes. Most of the recent modifications of this process have been targeted at an improved control of the morphology of nanotubes (including the length, diameter and size of agglomerates), a reduction in the synthesis temperature and process intensification.

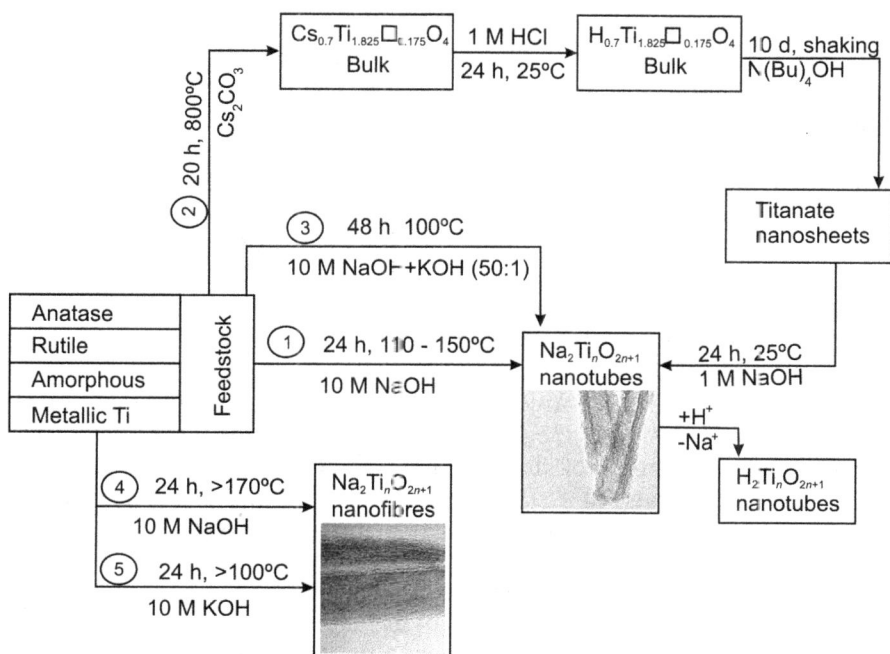

Figure 2.10 Prospective routes to the synthesis of titanate nanotubes and nanofibres.

Effective ways to control the length of nanotubes include: the ultrasonic treatment of initial raw TiO_2[44] or an improvement in the fluid flow and mass transport during alkaline hydrothermal treatment,[38] which probably improves the dynamics of nanotube growth in the axial direction due to the availability of the dissolved titanium(IV) species. The average diameter of nanotubes can be controlled, to some extent, by the synthesis temperature.[29] A degree of control over the shape of the nanotubular agglomerates can be achieved by using either hydrogen peroxide[45] or raw TiO_2 with a controlled initial particle size distribution.[44]

There are several approaches to process intensification of nanotubular titanate growth, including: microwave heating[46–49] or ultrasonication[50] of the reaction mixture during synthesis; improved mixing by use of a revolving autoclave;[51] or the hot press fabrication method.[52] All of these approaches allow the synthesis time to be reduced from 24 hours down to just a few hours.

The need to use pressurised reactors for the preparation of nanotubes greatly increases the manufacturing cost and complicates the health and safety requirements. Attempts to avoid autoclave operations, by a reduction of the synthesis temperature below the boiling temperature of the alkaline solution (*ca.* 106 °C), usually result in the formation of multilayered, lepidocrocite-type nanostructures (nanosheets) instead of nanotubes.[53,54] There are several reports of titanate nanotubes obtained under reflux conditions.[55] In such cases, local overheating of the reaction mixture is possible or a particular form of raw TiO_2 (characterised by a higher rate of dissolution in alkaline solvents) tends to be used.

One of the prospective low-temperature routes to titanate nanotubes is *via* titanate nanosheets, which can be produced by exfoliation of lepidocrocite-type caesium titanate (route 2 in Figure 2.10). The method is based on the phenomenon of the spontaneous formation of titanate nanotubes, at room temperature, with the addition of NaOH to a colloidal solution of titanate nanosheets.[56] The method includes: the calcination stage of TiO_2 with Cs_2CO_3; followed by ion-exchange of Cs^+ to H^+; and then exfoliation of titanate nanosheets in the presence of tetrabutylammonium hydroxide (TBAOH) at room temperature.[57] This route does not require the use of autoclaves and most of the stages can be undertaken under ambient conditions. However, the limitations of this method are an increased overall process time and the multi-stage nature of the process.

One approach adopted to reduce the temperature during the synthesis of titanate nanotubes, involves the search for a solvent, or mixture of the solvents, in which the concentration of dissolved Ti(IV) is similar at low temperatures, to that in pure NaOH at 110–150 °C. This has resulted in a lower temperature route to the synthesis of nanotubes using a mixture of NaOH with KOH[58] (route 3 in Figure 2.10), allowing a substantial conversion to be achieved at approximately 100 °C, under reflux conditions and atmospheric pressure. Further improvements to this route might include an optimisation of the ratio between NaOH and KOH to further lower the temperature, and the use of additives to achieve a high conversion of TiO_2 to titanate nanotubes within several hours under simple reflux conditions.

The formation of titanate nanofibres usually occurs during the hydrothermal treatment of TiO_2 raw materials with NaOH solution ($10 \, mol \, dm^{-3}$) at temperatures higher than 170 °C (route 4 in Figure 2.10).[29,59] The use of KOH ($10 \, mol \, dm^{-3}$) solution as a solvent also results in the formation of nanofibres,[60,62] while a mixture of nanotubes and nanofibres tend to be formed at lower temperatures[61,62] (route 5 in Figure 2.10).

2.3 Electrochemical (Anodic) Oxidation

Following the discovery of the hexagonal-packed porous aluminium oxide obtained by anodising aluminium,[63] many attempts have been made to understand the mechanism of ordered pore formation and the methods for controlling the porous structure and morphology of porous films, by tailoring the electrochemical conditions and the composition of electrolyte. In addition, attempts to translate this method to other valve metals have also been the subject of many investigations. This section describes the methods for anodised TiO_2 nanotube array synthesis; the mechanism of nanotube formation and morphology control are also considered.

2.3.1 Principles and Examples

In the past, most research in the area of titanium anodising was focussed on the preparation of the non-porous, durable and corrosion-resistant film of TiO_2. However, the recent developments in nanotechnology and the demand for new nanomaterials have stimulated the development of methods for preparing porous TiO_2 films. It is well known that the addition of fluoride ions to an aqueous electrolyte solution can significantly lower the corrosion resistance of titanium and a TiO_2 coating, due to the formation of pitting channels. Based on this effect, a new method to produce nanoporous TiO_2 films has been established using fluoride-containing electrolytes.

In early 2001, Grimes *et al.* reported the preparation of self-organised TiO_2 nanotube arrays by direct anodising of titanium foil in a H_2O–HF electrolyte[64] at room temperature. The nanotubes were all oriented in the same direction, perpendicular to the surface of the electrode, forming a continuous film. The thickness of this film (or the length of the tubes) was only 200 nm (see Table 2.2). The average internal diameter of nanotubes exceeded 50 nm (see Figure 2.11). One end of the nanotubes (facing the electrolyte) was always open, while the other end (which was in contact with the titanium electrode) was always closed (Figure 2.11c). The thickness of the walls of the nanotubes was approximately 10 nm. Usually the walls were thinner at the end of the nanotubes facing the electrolyte, and thicker at the end in contact with the titanium electrode (Figure 2.11d). In this preparation, the wall of nanotubes consists of amorphous TiO_2, which can be converted to polycrystalline anatase or an anatase–rutile mixture by heat treatments at temperatures above 500 °C. The surface of the

Table 2.2 Morphological properties of TiO$_2$ nanotubes produced by anodising Ti foil in various electrolytes at 25 °C.

Electrolyte composition	*Electrode potential vs SCE/V*	*Internal diameter/ nm*	*Length / μm*	*Ref.*
0.5–3.5 wt% HF in H$_2$O	3–20	25–65	0.2	64
0.5 wt% NH$_4$F in 1 mol dm^{-3} (NH$_4$)$_2$SO$_4$	20	90–110	0.5–0.8	65
4 wt% HF in 48 wt% DMSO, 48 wt% ethanol	20	60	2.3	66
0.5 wt% NH$_4$F in 1 mol dm^{-3} (NH$_4$)H$_2$PO$_4$, + 1 mol dm^{-3} H$_3$PO$_4$	20	40–100	0.1–4	67
0.5 wt% NH$_4$F in CH$_3$COOH	10–120	20	0.1–0.5	68
0.1–1 wt% NaF in 0.1–2 1 mol dm^{-3} Na$_2$SO$_4$	20	100	2.4	69
0.2 wt% H$_2$O in ethylene glycol with 0.2 mol dm^{-3} HF	120	70–200	260	70
H$_2$O–glycerol (from 50:50 to 0:100 vol.%), 0.27 mol dm^{-3} NH$_4$F	2–40	20–300	0.15–3	89
0.5 mol dm^{-3} HCl and 0.5 mol dm^{-3} H$_2$O$_2$ in H$_2$O + ethylene glycol	10–23	15	0.86	71

nanotube walls can be corrugated or smooth, depending on the electrolyte composition.

More recently, a number of fluoride ion-containing electrolytes, including: a NH$_4$F–(NH$_4$)$_2$SO$_4$ mixture,[65] HF in a dimethyl sulfoxide (DMSO)–ethanol mixture,[66] phosphate,[67] acetate,[68] a non-acidic Na$_2$SO$_4$–NaF mixture[69] and an electrolyte containing ethylene glycol,[70] have been used. Some attempts have also been made to use a fluoride-free electrolyte, based on a mixture of HCl with H$_2$O$_2$.[71] The morphological properties of the resultant nanotubes are summarised in Table 2.2. Several major review papers, which consider the fabrication, properties and various applications of anodised ordered TiO$_2$ nanotubular coatings have been published.[72–74]

In contrast to techniques using AAO as a template, TiO$_2$ nanotubes prepared by direct anodising are not usually separated from each other in a regular manner and do not have well-developed cavities between the tubes (see Figures 2.3. and 2.11). Usually, in such cases, the walls of aligned nanotubes are mutually connected by small bridges.

Starting from titanium alloys (titanium combined with other valve metals) instead of pure titanium, it is possible to prepare nanotube arrays of mixed oxide composition using the method of anodising in fluoride ion-containing electrolytes. Such composite oxide nanotubes can increase drastically the potential functionality of the tubes (*e.g.* the incorporation of doping species on

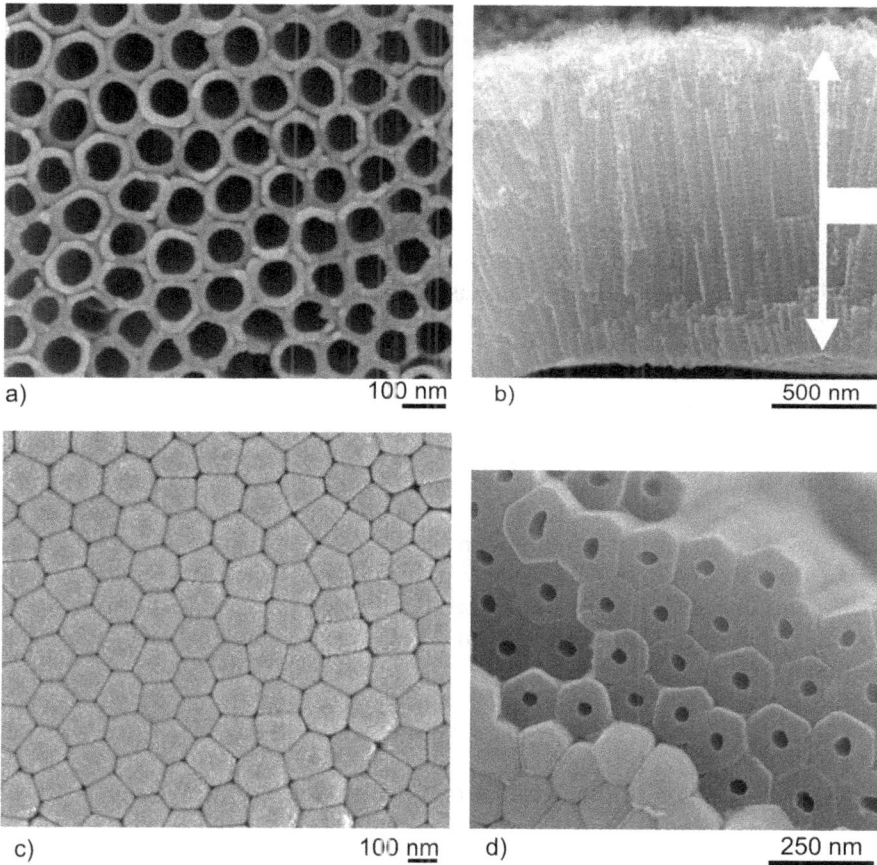

Figure 2.11 Typical SEM images of TiO$_2$ nanotube array films prepared by anodic oxidation of titanium in fluoride-containing electrolytes: a) top view and b) side view of the wall structure, and c) and d) bottom views. (Images are reproduced with kind permission as follows: a) and c) from ref. 72 b) from ref. 67 and d) from ref. 70).

the oxide structure), hence expanding the potential for industrial applications. Furthermore, novel nanostructured composites may also have interesting properties. A range of valve metal oxide nanotube arrays have been reported, including: binary TiAl (ref. 75), TiNb (ref. 76) and TiZr (ref. 77,78), as well as complex Ti$_6$Al$_7$Nb (ref. 79) and Ti$_{29}$Nb$_{13}$Ta$_{4.6}$Zr.(ref. 80). The morphological properties of mixed oxide nanotube arrays are similar to those for TiO$_2$ nanotubes. The range of metals which are capable of forming mixed oxide nanotubes *via* anodising of the corresponding titanium alloy is limited, due to the difference in solubility of each alloy component in the electrolyte, resulting in a selective dissolution of the least stable element and different reaction rates for the different alloy phases.[74]

The advantage of TiO_2 nanotubes produced by anodising is that they have been effectively immobilised on a titanium surface during preparation. As the result, these nanotubes have several possible applications. It has recently been found that the electroconductivity of TiO_2 nanotube films increases by several orders of magnitude in the presence of gaseous hydrogen at 290 °C (ref. 81), making this a promising material for hydrogen sensing. Similar TiO_2 nanotubes have also shown promise for use as photocatalytic, self-cleaning surfaces,[82] as photoanodes for water splitting[83] or in dye-sensitised solar cells[84–86] (where the efficiency of the photoanodic response depends on the nanotube wall thickness, and the current collection efficiency is a function of the quality of the nano-tube–electrode contact).

2.3.2   Mechanism of Nanotube Growth

For better control over the morphology and the degree of ordering in nano-tubes, it is vital to understand the underlying principles and mechanism for the formation of aligned nanotubes under anodic conditions. The growth of nanotubes by anodising titanium can be described as a selective etching, and the method can be related to a *top down* approach. In the simplest approach, such nanotube growth can be described in terms of a competition between several electrochemical and chemical reactions, including: anodic oxide for-mation:

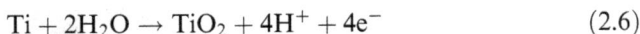

$$Ti + 2H_2O \rightarrow TiO_2 + 4H^+ + 4e^- \qquad (2.6)$$

chemical dissolution of the titanium oxide as soluble fluoride complexes, *e.g.*:

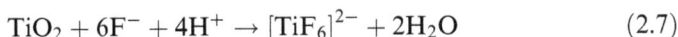

$$TiO_2 + 6F^- + 4H^+ \rightarrow [TiF_6]^{2-} + 2H_2O \qquad (2.7)$$

and direct complexation of Ti^{4+} ions migrating through the film:

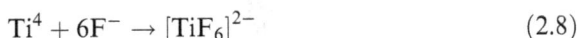

$$Ti^4 + 6F^- \rightarrow [TiF_6]^{2-} \qquad (2.8)$$

Reaction (2.6) describes the oxide growth on an anodized metal surface in a fluoride-free electrolyte. Firstly, a layer of anodically formed oxide is formed. Further oxide growth is controlled by the migration of O^{2-} and Ti^{4+} ions through the growing oxide film. As the system is under a constant applied voltage, the electric field within the oxide is progressively reduced by the increasing oxide thickness, the process is self-limiting. The question is why under certain conditions, does the formation of cylindrical pores organized in hexagonally symmetrical arrays occur?

In aqueous electrolytes and at constant potential, most valve metals give rise to current–time curves with an exponential decay shape, due to the passivation of the electrode surface as a result of the formation of a barrier layer of low-conductivity metal oxide (reaction 2.6 and Figure 2.12, top left hand side). In contrast, the addition of HF or another source of fluoride ions, may result in an

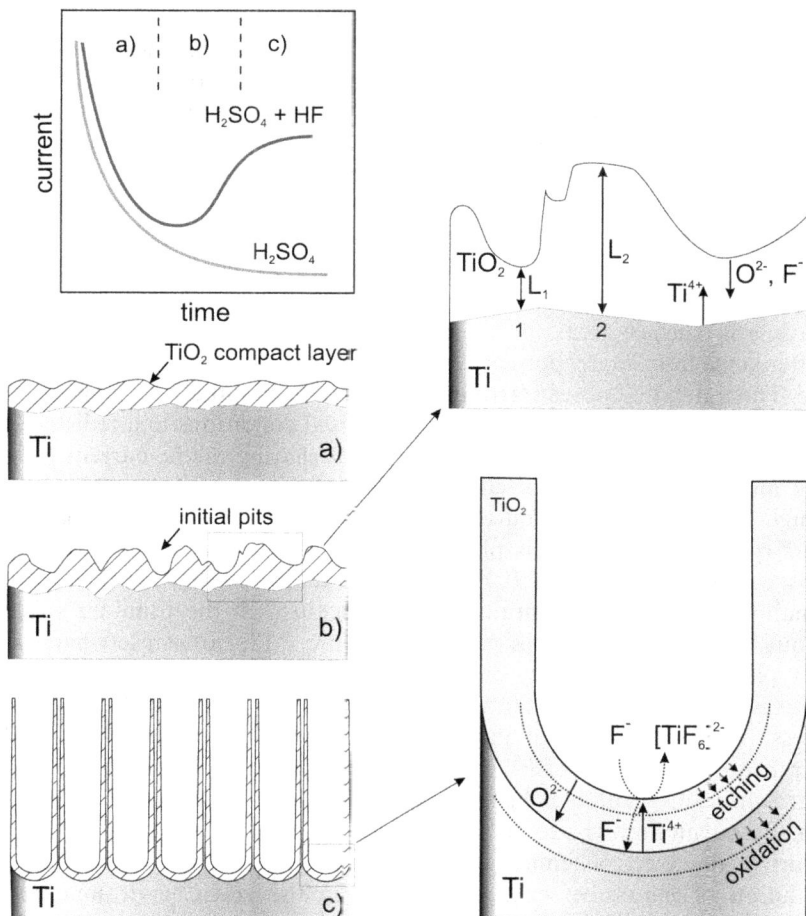

Figure 2.12 Schematic representation of the film formation of a TiO_2 nanotubular array under anodic conditions. Top left-hand side: typical current–time plot for the oxidation of Ti in an electrolyte with, or without, F^- ions. Diagrams a), b) and c) show the morphology of the coating during the corresponding phases of the process. Right-hand side: details of ion transport occurring in phases b) and c).

initial exponential decrease of current (phase a) followed by an increase (phase b) to the quasi steady-state level (phase c). The steady-state level and the rate of the current recovery are increased with an increase in fluoride concentration.[74]

Typically, such behaviour of the current can be ascribed to different stages in the pore formation process, as schematically illustrated in Figure 2.12 (where drawings a, b and c correspond to the phases a, b and c in the current–time curve for fluoride-containing electrolyte). In the first stage, a barrier oxide is formed, leading to a decay in current (phase a) due to the reduced electroconductivity of the layer.

Due to the roughness of the barrier layer and the barrier layer–metal interface, different parts of the film have different film thicknesses L_1 and L_2 (see Figure 2.12, right hand side at points 1 and 2). Such non-uniformity results in an uneven distribution of the electric field within the film, causing faster ion migration in the thinner areas (point 1) than in the thicker areas (point 2). This effect can be particular pronounced due to the high mobility of fluoride ions in TiO_2 film. The differences in ion transport result in differences in the local current densities and, as a consequence, in the local dissolution rates. During this stage, the surface is locally activated and pores start to grow randomly (phase b). This is usually accompanied by a rise in current due to an increase in the available surface area.

After some time, many pores have been initiated and a tree-like growth takes place. The individual pores start interfering with each other and competing for the available current. This leads under optimised conditions to a redistribution of local current density, resulting in an equal sharing of the current between pores, and accompanied by a self-assembling of the pores. The current passing through the electrode is stabilised and the steady-state growth of nanotubes occurs (phase c). During this phase, the rate of titanium oxide formation is almost equal to the rate of $[TiF_6]^{2-}$ formation and dissolution. In this situation, the nanotube oxide cap continuously moves through the titanium substrate without thickening nanotubes walls (see Figure 2.12, bottom left hand side). The typical current efficiency of TiO_2 nanotube formation in an acidic electrolyte is relatively low $(3–10\%)$[74] and is still decreasing at the end of the process. The overall rate of the process in steady-state phase is limited by the transport (diffusion) of F^- inside the channel from bulk solution towards the growing TiO_2 cap, and the transport of $[TiF_6]^{2-}$ in the opposite direction. Both effects can limit the total current.

During the electrochemical growth of TiO_2 nanotubes, slow chemical dissolution of nanotube walls also occurs in the acidic environment. This results in the nanotube end which faces the electrolyte having a thinner wall compared with that of the nanotube end facing the substrate, as can be seen by comparing Figure 2.10a and Figure 2.10d. This is a consequence of the variation in electrolyte exposure time of the different nanotube ends.[87] However, the reason for the separation of pores into tubes (unlike the case of AAO) is not yet clear.

Since the difference in densities of TiO_2 from metal titanium is significant, during TiO_2 growth there is an internal stress, which pushes the metal oxide up, resulting in an increase in nanotube length of approximately 20%.[87]

2.3.3 Methods to Control the Morphology of Nanotubes

For practical purposes, it is important to know what methods exist for adjusting the morphological properties of anodic TiO_2 nanotubes, including: tube diameter and length, wall thickness and roughness, as well as the degree of ordering of the aligned nanotubes.

Control of nanotube diameter. Anodic oxidation of titanium in fluoride-free electrolyte usually leads to the growth of a compact layer of TiO_2. The thickness of the layer is linearly proportional to the applied potential to the electrode, and the proportionality coefficient is in range of 1 to 5 nm V^{-1} (ref. 88). In contrast, the increase in electrode potential during the growth of TiO_2 nanotubes in fluoride-containing electrolytes, usually results in an increase in the nanotube diameter. This has been confirmed, for example, by SEM images of TiO_2 nanotubes grown in a water–glycerol electrolyte[89] at various applied potentials (see Figure 2.13, top). The functional dependence of the average nanotube diameter on applied potential is linear across a wide range of applied potential (up to 40 V). Figure 2.13 (bottom) shows the change in nanotube diameter with increasing potential in electrolytes of differing composition,

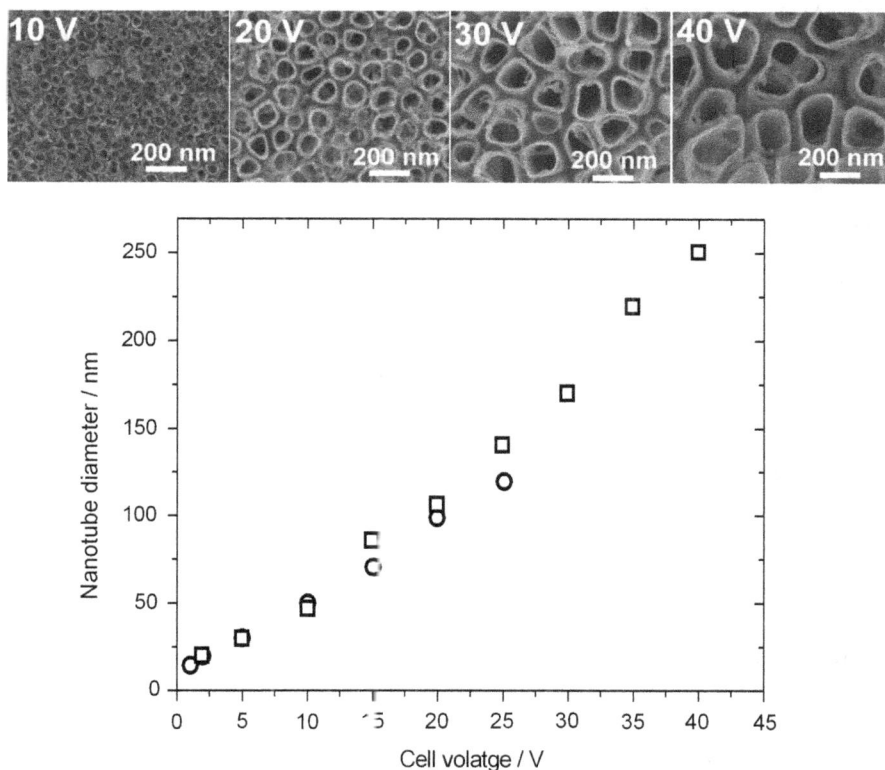

Figure 2.13 Top: SEM images of TiO_2 nanotubes grown in glycerol–water (50:50 vol.%) electrolyte containing NH_4F (0.27 mol dm^{-3}) at various applied potentials. (Images are reproduced with kind permission from ref. 89). Bottom: TiO_2 nanotube internal diameters shown as a function of the applied potential during anodising in aqueous electrolyte containing ($\square$) glycerol–water (50:50 vol.%) with NH_4F (0.27 mol dm^{-3}; ref. 89), and ($\bigcirc$) H_3PO_4 (1 mol dm^{-3}) with 0.3 wt% HF (ref. 90).

including a phosphoric acid electrolyte.[90] Nanotube diameters of up to 250 nm can be achieved without a significant deterioration in the ordering of nanotubes. A similar dependence of nanotube diameter on applied voltage is also true for titanium alloys.[91]

Control of nanotube lengths. The length of anodised TiO_2 nanotubes is determined by the ratio between the rates of pore growth from the electrochemical oxidation-etching of titanium substrate, and the nanotube top end dissolution resulting from chemical reaction with electrolyte components. The latter can be decreased by the use of less corrosive electrolytes [*e.g.*, NaF (ref. 69) or NH_4F (ref. 88) as a source of fluoride ions instead of HF], which can prevent the rapid dissolution of nanotube walls. The rate of pore electrochemical etching can be facilitated by applying conditions which favour a thinner barrier layer. A thinner barrier layer may result in faster ion transport, allowing an increased current density, leading to a higher rate of nanotube growth and a shorter exposure time to aggressive electrolyte. A recent approach uses polar organic electrolytes with minimal water content, which can inhibit the donation of oxygen atoms, decrease the tendency to form an oxide, and reduce the thickness or lower the quality of the barrier layer. A thinner barrier layer can enhance the transport of ions.[89] The use of quaternary ammonium ions can also inhibit the formation of a thick barrier layer.[92]

If the rate of titanium substrate dissolution is rapid during nanotube growth, then the length of the nanotubes becomes limited by the transport of fluoride ions inside their channels. In general, the longer the tube, the slower the ion transport in the channel. When the nanotube length exceeds a certain value, the overall rate of nanotube growth becomes equal to the rate of their dissolution. This determines the length of the nanotubes. The rate of diffusion can be increased to some extent by increasing the fluoride ion concentration in the electrolyte, by increasing the gradient of concentration between the bulk solution and the surface of the electrode. This usually results in an increase in nanotube length.[72]

Control of nanotube ordering. Under certain conditions, anodic TiO_2 nanotubes are assembled into an array with a high degree of hexagonal ordering. Several factors strongly influence the degree of ordering:[93] the anodising cell voltage (the highest possible voltage just below dielectric breakdown appears to be the optimal voltage), and the purity of the material (certain ordering faults can be eliminated by using a high-purity Ti). In addition, repeated anodising (as in the case of Al) can clearly improve ordering. By using this approach, the bottom imprints of a first tube layer in the underneath Ti act as "pre-ordering" guides for subsequent anodic tube initiation and growth – an *in situ* templating effect.

Control of wall thickness. The typical thickness of the TiO_2 nanotubes wall is a few tens of nanometres. Despite the importance of this parameter for some practical applications including hydrogen sensing,[72] there is only limited data available regarding the control of nanotube wall thickness. One possible approach would be to use the recently observed correlation between nanotube wall thickness and the concentration of fluoride ions in solution; an increase in fluoride ion concentration can result in a decrease in nanotube wall thickness.[94]

The type of organic additive in the electrolyte may also affect the smoothness of the nanotube walls. For example, the addition of acetic acid to electrolyte can significantly decrease the roughness of nanotube walls.[68] A decrease in the water content of a glycerol electrolyte also increases the smoothness of nanotube walls.[89]

At the present time, by adjusting the anodising conditions (including: cell voltage, reaction time, electrolyte composition, addition of additives or the use of an organic electrolyte), it is possible to control the internal diameter of nanotubes from 12 nm to 242 nm, wall thickness from 5 nm to 34 nm (ref. 72) and the length of nanotubes up to 260 μm.[70]

Conclusions

In this chapter, three different approaches for the preparation of elongated TiO_2 nanostructures have been reviewed, namely template-assisted, alkaline hydrothermal and electrochemical anodising methods. The first two methods represent a *bottom up* approach, whereas the third one corresponds to the *top down* fabrication method. All three methods allow the synthesis of nanostructures having a wide range of morphological properties and dimensions. Each of the three methods is characterised by an individual mechanism of nanotube formation, and as a consequence they all have different principles governing the morphology control of nanostructure formation. Only titanate and TiO_2 nanostructures are considered in this chapter, but it is possible that similar principles and methods may be applied to the preparation of novel non-titanium oxide based nanostructures.

References

1. J. C. Hulteen and C. R. Martin, *J. Mater. Chem.*, 1997, **7**, 1075.
2. C. Bae, H. Yoo, S. Kim, K. Lee, J. Kim, M. M. Sung and H. Shin, *Chem. Mater.*, 2008, **20**, 756.
3. G. L. Hornyak, H. F. Tibbals, J. Dutta and J. J. Moore, *Introduction to Nanoscience and Nanotechnology*, CRC Press, Taylor and Francis Group, Boca Raton, London, New York, 2009.
4. C. Dionigi, P. Greco, G. Ruani, M. Cavallini, F. Borgatti and F. Biscarini, *Chem. Mater.*, 2008, **20**, 7110.
5. S. Kobayashi, K. Hanabusa, N. Hamasaki, M. Kimura, H. Shirai and S. Shinkai, *Chem. Mater.*, 2000, **12**, 1523.
6. B. F. Cottam and M. S. P. Shaffer, *Chem. Commun.*, 2007, 4378.
7. J. Y. Gong, S. R. Guo, H. S. Qian, W. H. Xu and S. H. Yu, *J. Mater. Chem.*, 2009, **19**, 1037.
8. S. Zhan, D. Chen, X. Jiao and C. Tao, *J. Phys. Chem. B*, 2006, **110**, 11199.
9. W. Y. Gan, H. Zhao and R. Amal, *Appl. Catal. A*, 2009, **354**, 8.

10. J. H. Jung, H. Kobayashi, K. J. C. Bommel, S. Shinkai and T. Shimizu, *Chem. Mater.*, 2002, **14**, 1445.
11. G. Gundiah, S. Mukhopadhyay, U. G. Tumkurkar, A. Govindaraj, U. Maitrab and C. N. R. Rao, *J. Mater. Chem.*, 2003, **13**, 2118.
12. T. Peng, A. Hasegawa, J. Qiu and K. Hirao, *Chem. Mater.*, 2003, **15**, 2011.
13. S. Fujikawa and T. Kunitake, *Langmuir*, 2003, **19**, 6545.
14. C. Hippe, M. Wark, E. Lork and G. Schulz-Ekloff, *Microporous Mesoporous Mater.*, 1999, **31**, 235.
15. Y. Qiu and J. Yu, *Solid State Commun.*, 2008, **148**, 556.
16. A. Michailowski, D. Al-Mawlawi, G. S. Cheng and M. Moskovits, *Chem. Phys. Lett.*, 2001, **349**, 1.
17. S. Z. Chu, K. Wada, S. Inoue and S. I. Todoroki, *Chem. Mater.*, 2002, **14**, 266.
18. S. M. Liu, L. M. Gan, L. H. Liu, W. D. Zhang and H. C. Zeng, *Chem. Mater.*, 2002, **14**, 1391.
19. P. Hoyer, *Langmuir*, 1996, **12**, 1411.
20. M. Adachi, Y. Murata, I. Okada and S. Yoshikawa, *J. Electrochem. Soc.*, 2003, **150**(8), G488.
21. S. Ngamsinlapasathian, S. Sakulkhaemaruethai, S. Pavasupree, A. Kitiyanan, T. Sreethawong, Y. Suzuki and S. Yoshikawa, *J. Photochem. Photobiol., A*, 2004, **164**, 145.
22. T. Kasuga, M. Hiramatsu, A. Hoson, T. Sekino and K. Niihara, *Langmuir*, 1998, **14**, 3160.
23. T. Kasuga, M. Hiramatsu, A. Hoson, T. Sekino and K. Niihara, *Adv. Mater.*, 1999, **11**, 1307.
24. Y. Lan, X. Gao, H. Zhu, Z. Zheng, T. Yan, F. Wu, S. P. Ringer and D. Song, *Adv. Funct. Mater.*, 2005, **15**, 1310.
25. X. D. Meng, D. Z. Wang, J. H. Liu and S. Y. Zhang, *Mater. Res. Bull.*, 2004, **39**, 2163.
26. C. C. Tsai and H. Teng, *Chem. Mater.*, 2006, **18**, 367.
27. Q. Chen, W. Zhou, G. Du and L. M. Peng, *Adv. Mater*, 2002, **14**, 1208.
28. M. Zhang, Z. Jin, J. Zhang, X. Guo, J. Yang, W. Li, X. Wang and Z. Zhang, *J. Mol. Catal. A: Chem.*, 2004, **217**, 203.
29. D. V. Bavykin, V. N. Parmon, A. A. Lapkin and F. C. Walsh, *J. Mater. Chem.*, 2004, **14**, 3370.
30. R. Ma, Y. Bando and T. Sasaki, *J. Phys. Chem. B*, 2004, **108**, 2115.
31. S. Zhang, L. M. Peng, Q. Chen, G. H. Du, G. Dawson and W. Z. Zhou, *Phys. Rev. Lett.*, 2003, **91**(25), 256103.
32. S. Zhang, Q. Chen and L. M. Peng, *Phys. Rev. B*, 2005, **71**, 014104.
33. C. C. Tsai and H. Teng, *Chem. Mater.*, 2004, **16**, 4352.
34. M. J. Paek, H. W. Ha, T. W. Kim, S. J. Moon, J. O. Baeg, J. H. Choy and S. J. Hwang, *J. Phys. Chem. C*, 2008, **112**, 15966.
35. W. Sugimoto, O. Terabayashi, Y. Murakami and Y. Takasu, *J. Mater. Chem.*, 2002, **12**, 3814.
36. G. Mogilevsky, Q. Chen, A. Kleinhammes and Y. Wu, *Chem. Phys. Lett.*, 2008, **460**, 517.

37. K. Yada, *Acta Crystallogr.*, 1971, **A27**, 659.
38. A. Kukovecz, M. Hodos, E. Horvath, G. Radnoczi, Z. Konya and I. Kiricsi, *J. Phys. Chem. B*, 2005, **109**, 17781.
39. M. Wei, Y. Konishi, H. Zhou, H. Sugihara and H. Arakawa, *Solid State Commun.*, 2005, **133**, 493.
40. A. Thorne, A. Kruth, D. Tunstall, J. T. S. Irvine and W. Zhou, *J. Phys. Chem., B*, 2005, **109**, 5439.
41. A. Kulak, D. V. Bavykin and F. C. Walsh, in preparation.
42. C. C. Chung, T. W. Chung and T. C. K. Yang, *Ind. Eng. Chem. Res.*, 2008, **47**, 2301.
43. H. G. Yang and H. C. Zeng, *J. Am. Chem. Soc.*, 2005, **127**, 270.
44. N. Viriyaempikul, N. Saro, T. Charinpanitkul, T. Kikuchi and W. Tanthapanichakoon, *Nanotechnology*, 2008, **19**, 035601.
45. Y. Mao, M. Kanungo, T. Hemraj-Benny and S. S. Wong, *J. Phys. Chem. B*, 2006, **110**, 702.
46. Y. Wang, J. Yang, J. Zhang. H. Liu and Z. Zhang, *Chem. Lett.*, 2005, **34**(8), 1168.
47. H. H. Ou, S. L. Lo and Y. H. Liou, *Nanotechnology*, 2007, **18**, 175702.
48. X. Wu, Q. Z. Jiang, Z. F. Ma, M. Fu and W. F. Shangguan, *Solid State Commun.*, 2005, **136**, 513.
49. X. Wu, Q. Z. Jiang, Z. F. Ma and W. F. Shangguan, *Solid State Commun.*, 2007, **143**, 343.
50. Y. Ma, Y. Lin, X. Xiao, X. Zhou and X. Li, *Mater. Res Bull.*, 2006, **41**, 237.
51. E. Horvath, A. A. Kukovecz, Z. Konya and I. Kiricsi, *Chem. Mater.*, 2007, **19**, 927.
52. T. Kubo, A. Nakahira and Y. Yamasaki, *J. Mater. Res.*, 2007, **22**(5), 1286.
53. M. Wei, Y. Konishi and H. Arakawa, *J. Mater. Sci.*, 2007, **42**, 529.
54. H. K. Seo, G. S. Kim, S. G. Ansari, Y. S. Kim, H. S. Shin, K. H. Shim and E. K. Suh, *Solar Energy Mater. Solar Cells*, 2008, **92**, 1533.
55. J. J. Yang, Z. S. Jin, X. D. Wang, W. Li, J. W. Zhang, S. L. Zhang, X. Y. Guo and Z. J. Zhang, *Dalton Trans.*, 2003, **20**, 3898.
56. R. Ma, Y. Bando and T. Sasaki, *J Phys. Chem. B*, 2004, **108**, 2115.
57. N. Sakai, Y. Ebina, K. Takada and T. Sasaki, *J. Am. Chem. Soc.*, 2004, **126**, 5851.
58. D. V. Bavykin, B. A. Cressey, M. E. Light and F. C. Walsh, *Nanotechnology*, 2008, **19**, 275604.
59. Z. Y. Yuan and B. L. Su, *Colloids Surf. A*, 2004, **241**, 173.
60. G. H. Du, Q. Chen, P. D. Han, Y. Yu and L. M. Peng, *Phys. Rev. B*, 2003, **67**, 035323.
61. D. V. Bavykin and, B. A. Cressey and F. C. Walsh, *Aust. J. Chem.*, 2007, **60**, 95.
62. W. A. Daoud and G. K. H. Pang, *J. Phys. Chem. B*, 2006, **110**, 25746.
63. H. Masuda and K. Fukuda. *Science*, 1995, **268**, 1466.
64. D. Gong, C. A. Grimes, O. K. Varghese, W. Hu, R. S. Singh, Z. Chen and E. C. Dickey, *J. Mater. Res.*, 2001, **16**, 3331.

65. L. V. Taveira, J. M. Macak, H. Tsuchiya, L. F. P. Dick and P. Schmuki, *J. Electrochem. Soc.*, 2005, **152**, B405.
66. C. Ruan, M. Paulose, O. K. Varghese, G. K. Mor and C. A. Grimes, *J. Phys. Chem. B*, 2005, **109**, 15754.
67. A. Ghicov, H. Tsuchiya, J. M. Macak and P. Schmuki, *Electrochem. Commun.*, 2005, **7**, 505.
68. H. Tsuchiya, J. M. Macak, L. Taveira, E. Balaur, A. Ghicov, K. Sirotna and P. Schmuki, *Electrochem. Commun.*, 2005, **7**, 576.
69. J. M. Macak, K. Sirotna and P. Schmuki, *Electrochim. Acta*, 2005, **50**, 3679.
70. S. P. Albu, A. Ghicov, J. M. Macak and P. Schmuki, *Phys. Status Solidi (RRL)*, 2007, **1**, R65.
71. N. K. Allam, K. Shankar and C. A. Grimes, *J. Mater. Chem.*, 2008, **18**, 2341.
72. C. A. Grimes, *J. Mater. Chem.*, 2007, **17**, 1451.
73. G. K. Mor, O. K. Varghese, M. Paulose, K. Shankar and C. A. Grimes, *Sol. Energy Mater. Sol. Cells*, 2006, **90**, 2011.
74. J. M. Macak, H. Tsuchiya, A. Ghicov, K. Yasuda, R. Hahn, S. Bauer and P. Schmuki, *Curr. Opin. Solid State Mater. Sci.*, 2007, **11**, 3.
75. H. Tsuchiya, S. Berger, J. M. Macak, A. Ghicov and P. Schmuki, *Electrochem. Commun.*, 2007, **9**, 2397.
76. A. Ghicov, S. Aldabergerova, H. Tsuchiya and P. Schmuki, *Angew. Chem. Int. Ed.*, 2006, **45**, 6993.
77. K. Yasuda and P. Schmuki, *Adv. Mater.*, 2007, **19**, 1757.
78. K. Yasuda and P. Schmuki, *Electrochim. Acta*, 2007, **52**, 4053.
79. J. M. Macak, H. Tsuchiya, L. Taveira, A. Ghicov and P. Schmuki, *J. Biomed. Mater. Res.*, 2005, **75A**, 928.
80. H. Tsuchiya, J. M. Macak, A. Ghicov, Y. C. Tang, S. Fujimoto, M. Niinomi, T. Noda and P. Schmuki, *Electrochim. Acta*, 2006, **52**, 94.
81. O. K. Vardhese, D. Gong, M. Paulose, K. O. Ong, E. C. Dickey and C. A. Grimes, *Adv. Mater.*, 2003, **15**, 624.
82. G. K. Mor, M. A. Carvalho, O. K. Varghese, M. V. Pishko and C. A. Grimes, *J. Mater. Res.*, 2004, **19**(2), 628.
83. G. K. Mor, K. Shankar, M. Paulose, O. K. Varghese and C. A. Grimes, *Nano Lett.*, 2005, **5**(1), 191.
84. M. Paulose, K. Shankar, O. K. Varghese, G. K. Mor, B. Hardin and C. A. Grimes, *Nanotechnology*, 2006, **17**, 1446.
85. H. Wang, C. T. Yip, K. Y. Cheng, A. B. Djurisic, M. H. Xie, Y. H. Leung and W. K. Chan, *Appl. Phys. Lett.*, 2006, **89**, 023508.
86. K. Zhu, N. R. Neale, A. Miedaner and A. J. Frank, *Nano. Lett.*, 2007, **7**, 69.
87. K. Yasuda, J. M. Macak, S. Berger, A. Ghicov and P. Schmuki, *J. Electrochem. Soc.*, 2007, **154**, C472.
88. J. W. Schultze and M. M. Lohrengel, *Electrochim. Acta*, 2000, **45**, 2499.
89. J. M. Macak, H. Hildebrand, U. Marten-Jahns and P. Schmuki, *J. Electroanal. Chem.*, 2008, **621**, 254.

90. S. Bauer, S. Kleber and P. Schmuki, *Electrochem. Commun.*, 2006, **8**, 1321.
91. A. Ghicov, S. Aldabergerova, H. Tsuchiya and P. Schmuki, *Angew. Chem. Int. Ed.*, 2006, **45**, 6993.
92. K. Shankar, G. K. Mor, A. Fitzgerald and C. A. Grimes, *J. Phys. Chem. C*, 2007, **111**, 21.
93. J. M. Macak, S. P. Albu and P. Schmuki, *Phys. Status Solidi (RRL)*, 2007, **1**, R181.
94. A. Elsanousi, J. Zhang, H. M. H. Fadlalla, F. Zhang, H. Wang, X. Ding, Z. Huang and C. Tang, *J. Mater. Sci.*, 2008, **43**, 7219.

CHAPTER 3

Structural and Physical Properties of Elongated TiO$_2$ and Titanate Nanostructures

3.1 Crystallography

Synthetic nanostructured titanate and TiO$_2$ materials often have characteristic XRD patterns. Our understanding of the crystal structure of nanotubular titanates and TiO$_2$ is incomplete due to several difficulties. Firstly, there are a number of crystal modifications not only for pure titanium dioxide (anatase, rutile and brookite), but also for its protonated forms such as the polytitanic acids H$_{2m}$Ti$_n$O$_{2n+m}$. Secondly, the small size of the crystals leads to a small value of the coherence area, which results in a broadening of the reflections in the XRD pattern. Thirdly, wrapping along a certain crystallographic axis during the formation of nanotubes can result in the widening of peaks of a given Miller index, making interpretation and assignment of peaks more difficult. Another difficulty is that the nanotubular structure of titanates is relatively unstable and can undergo further phase transformation during heating, acid treatment, or other chemical treatments during, or after, preparation of nanotubes. The low weight of hydrogen atoms also results in difficulties in locating their precise positions and population inside the crystals. All of these factors contribute to a controversy over the exact crystal structure of titanate nanotubes. In this chapter, the possible crystal structures of titanate nanotubes and other nanostructures currently found in the literature are reviewed.

3.1.1 Crystallography of Titanate Nanotubes

In an early study, Kasuga *et al.*[1] characterized their product as anatase. In a recent paper, the crystal structure of nanotubular titanates was still been

RSC Nanoscience & Nanotechnology No. 12
Titanate and Titania Nanotubes: Synthesis, Properties and Applications
By Dmitry V. Bavykin and Frank C. Walsh
© Dmitry V. Bavykin and Frank C. Walsh 2010
Published by the Royal Society of Chemistry, www.rsc.org

interpreted as anatase,[2] despite the fact that the crystal structure of these nanotubes is known to be more sophisticated.

Based on XRD, SAED and HRTEM data, Peng et al.[3,4] proposed that the crystal structure of titanate nanotubes corresponded to the layered trititanic acid (H$_2$Ti$_3$O$_7$) having a monoclinic crystal structure (see Table 3.1). A schematic showing the crystal structure of monoclinic trititanic acid in a TiO$_6$ edge-sharing octahedron representation is shown in Figure 3.1. The three different projections of the crystal structure corresponding to crystallographic axes are linked to particular areas in HRTEM image of the nanotube, showing how the packaging of layers in the crystal is related to the fringes in the microscope image. A nanotubular morphology of layered trititanic acid can be obtained by rolling several (100) planes around axis [010] (ref. 3) such that the axis of the nanotube is parallel to the axis b of monoclinic H$_2$Ti$_3$O$_7$. The radial direction from the tube centre towards the walls corresponds to axis a of trititanic acid, and the tangential direction to the nanotube surface corresponds to axis c.

Recently, Wu et al.[5] have proposed that rolling of the (100) plane could occur around axis [001]. In both cases, the walls of the nanotubes consist of several layers, typically separated by 0.72 nm. The structure of each layer corresponds to the structure of the (100) plane of monoclinic titanates, which is a set of closely packed TiO$_6$, edge-sharing octahedra (see Figure 3.1).

Using XRD and TEM data, Nakahira et al.[6] characterized a TiO$_2$ nanotube sample produced by alkaline hydrothermal treatment, as a tetratitanic acid (H$_2$Ti$_4$O$_9 \cdot$ H$_2$O). The presented XRD spectra, however, have very broad reflections which have similar positions to those of trititanic acid. The crystal structure of tetratitanic acid is similar to trititanic acid, the former having four-edge sharing octahedra in the unit cell, rather than the three-edge sharing octahedra of the latter.

Based on studies of the sodium content of titanate nanotubes at various pH values during acid washing (in combination with XRD and TEM data), Jin et al.[7] proposed a crystal structure for titanate nanotubes as H$_2$Ti$_2$O$_4$(OH)$_2$ with an orthorhombic unit cell. Both protons of the bititanic acid could be ion-exchanged with sodium ions. The XRD pattern shows similar reflections to the monoclinic trititanic acid (see Table 3.1).

The orthorhombic dititanic acid also has a layered structure of walls, where each layer represents the (100) plane rolled around the b axis. The (100) plane is built up from edge-sharing TiO$_6$ octahedra, forming a zigzag structure (see Figure 3.2). After scrolling of nanosheets into nanotubes, the radial direction from the tube centre towards the walls of the tube corresponds to axis a of orthorhombic dititanic acid, and the tangential direction to the nanotube surface corresponds to axis c. The axis of the nanotubes is parallel to axis b, similar to that of trititanic acid.

Ma et al.[8,9] obtained XRD, Raman spectroscopy, X-ray absorption fine structure and electron diffraction data and suggested that lepidocrocite-type structures (H$_{0.7}$Ti$_{1.827}\square_{0.175}$O$_{4.0} \cdot$ H$_2$O, where $\square$ represents a vacancy), could be present in a mixture of layered nanotube titanates. Detailed electron diffraction studies of selected zones on the nanotubes also support this

Table 3.1 Comparison of the crystal structures of layered titanates.

Crystallographic phase	Symmetry	Lattice parameters/nm				XRD reflection, 2Θ/degree										Ref.
		a	b	c	β											
Nanotubes																
$H_2Ti_3O_7$	Monoclinic	1.602	0.375	0.919	101.5°	10.5		24.4	28	34	38.5	44.5	48.2		61.5	4
$H_2Ti_2O_4(OH)_2$	Orthorhombic	1.926	0.378	0.300	90°	11		24.4	29	33	38		48.4	60	62	7
$H_2Ti_4O_9 \cdot H_2O$	Monoclinic	1.825	0.379	1.201	106.4°	9		24.3	28	34	38		48		62	6
$H_xTi_{2-x/4}\square_{x/4}O_4 \cdot H_2O$	Orthorhombic	0.378	1.834	0.298	90°	10		24	28				48		62	8
TiO_2-B	Monoclinic	1.218	0.374	0.652	107.1°	9.5	15	25	29.5			44	48	57	62	13
Nanofibres																
$H_2Ti_5O_{11} \cdot H_2O$	Monoclinic	2.001	0.376	1.499	124.0°	10	14			36		43	46			17

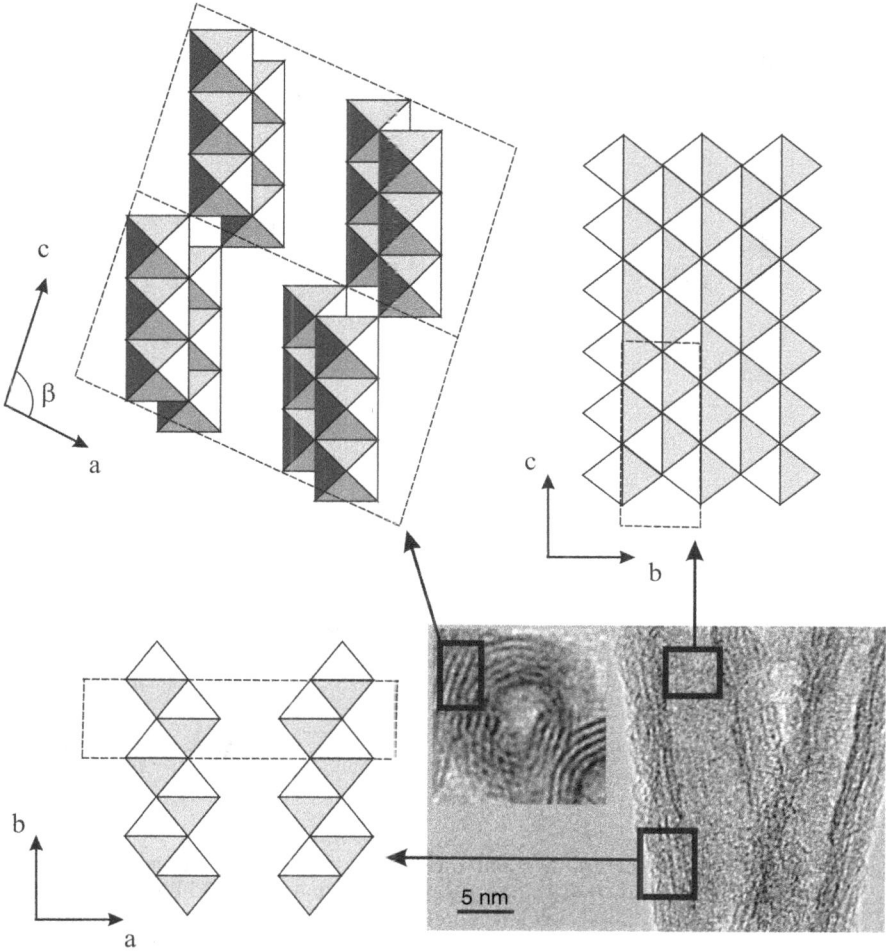

Figure 3.1 Crystal structure of trititanic acid (H$_2$Ti$_3$O$_7$) in three different projections, in octahedral presentation and a HRTEM image of titanate nanotubes. The solid rectangles in the HRTEM image show the areas corresponding to particular projections. The dashed lines show the dimensions of the unit cell. Hydrogen atoms are not shown, for clarity. a, b and c are crystallographic axes, and β is the monoclinic angle.

hypothesis.[10] Unlike trititanic acid, which has three steps of corrugated layers, the lepidocrocite titanate and orthorhombic dititanate consist of a continuous and planar two-dimensional array built up from TiO$_6$ edge-sharing octahedral similar to dititanic acid (see Figure 3.2), with the exception that axes a and b are swapped. Thus the radial direction from the tube centre towards the walls of the tube corresponds to axis b of lepidocrocite-type titanic acid.

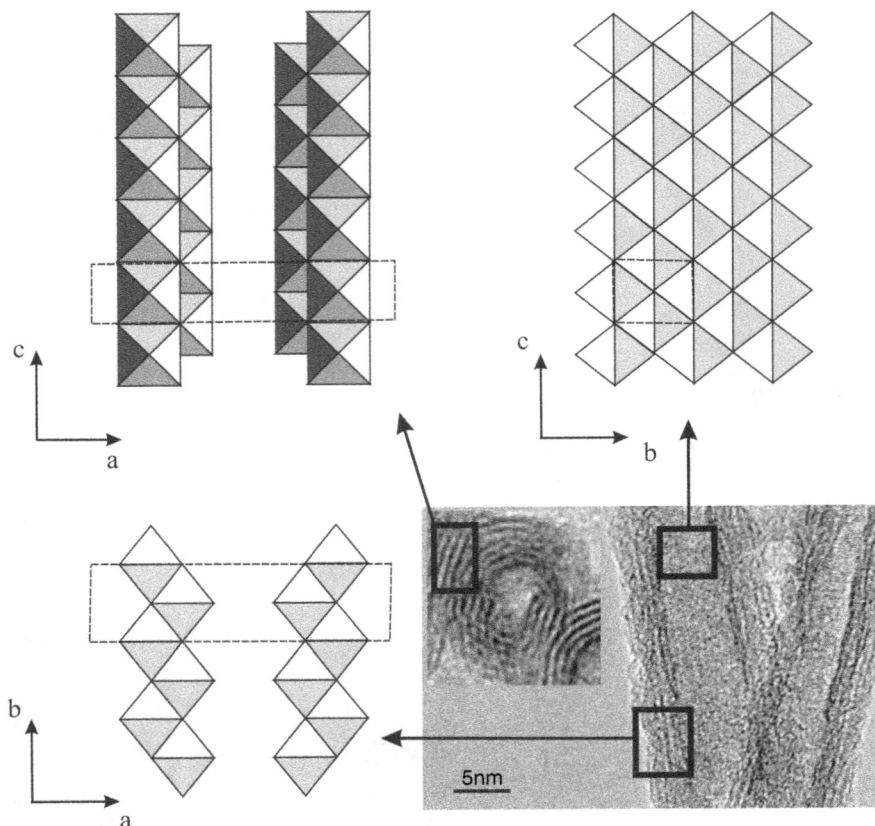

Figure 3.2   Crystal structure of titanic acid ($H_2Ti_2O_4(OH)_2$) in three different projections, shown in octahedral presentation and a HRTEM image of titanate nanotubes. Solid rectangles in the HRTEM image show the areas corresponding to particular projections. The dashed lines show the dimensions of unit cell. Atoms of hydrogen are not shown for clarity. a, b and c are crystallographic axes.

Another difference between monoclinic tri- or tetra-titanates and orthorhombic titanates is that after scrolling, the nanotube surface is corrugated in the first instances and smooth in the latter. The STM and AFM images reported elsewhere[11] showed the steps in the nanotube wall with their edges parallel to the tube axis supporting a monoclinic crystal structure of the nanotubes.

It was demonstrated that bulk-layered protonated poly-titanates could be transformed to the metastable monoclinic modification of titanium dioxide (TiO_2-B) under calcination.[12] This modification of titanium dioxide has a lower density than anatase or rutile and has a monoclinic unit cell. The structure of TiO_2-B is characterized by a combination of edge- and corner-sharing TiO_6

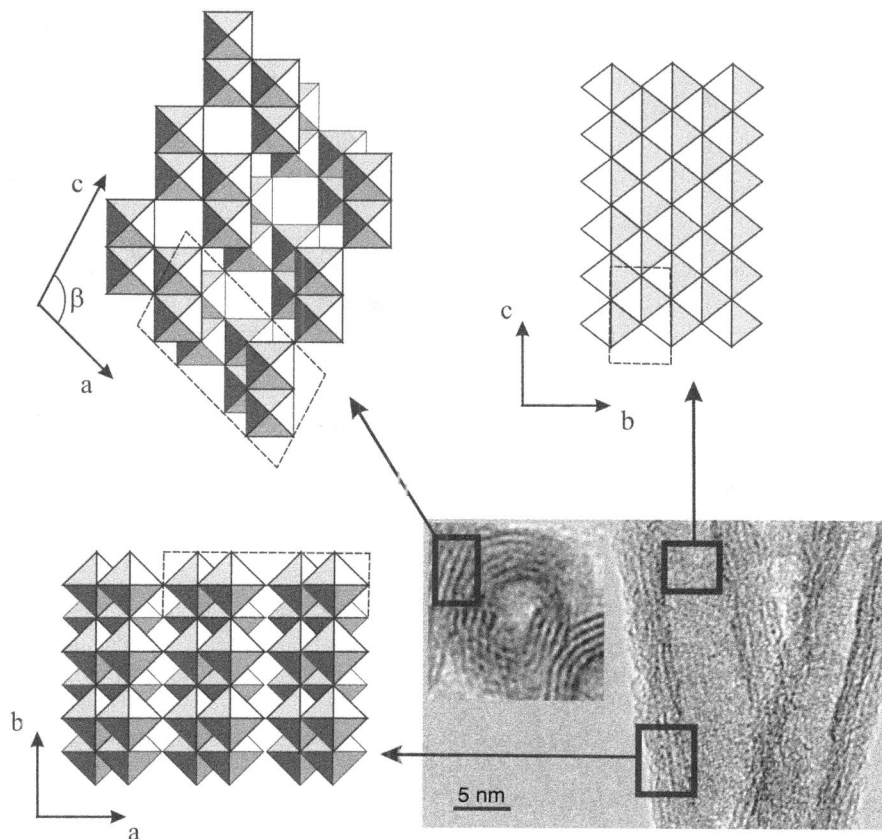

Figure 3.3 Crystal structure of monoclinic TiO$_2$-(B) in three different projections, in octahedral presentation and a HRTEM image of titanate nanotubes. The solid rectangles in the HRTEM image show the areas corresponding to particular projections. The dashed lines show the dimensions of a unit cell. Atoms of hydrogen are not shown for clarity. a, b and c are crystallographic axes, and β is the monoclinic angle.

octahedra, forming a structure with channels in which transport and exchange of small cations can occur (see Figure 3.3).

On the basis of XRD, TEM and Li$^+$ exchange studies, Bruce *et al.*[13] have suggested that the crystal structure of TiO$_2$ nanotubes produced by alkaline hydrothermal treatment, followed by acid washing and calcination, corresponds to the structure of TiO$_2$-B. Indeed, washing of sodium titanate nanotubes with acid or water results in the formation of the protonated form of titanate nanotubes. Following drying at increased temperatures, the dehydration of solids and the formation of TiO$_2$-B nanotubes could occur, resulting in the product having a density in the range of 3.64–3.76 g cm^{-3} (ref. 14), which is less than the density of anatase or rutile (3.9 and 4.25 g cm^{-3}, respectively). The XRD pattern of TiO$_2$-B crystals, however, is slightly different from that of TiO_2 nanotubes, especially at small 2θ values (see Table 3.1). As our understanding of the exact structure of titanate nanotubes continues to evolve, it is common to see both 'titanate nanotubes' and 'TiO_2 nanotubes' used as terms to describe these materials.

A recent analysis of pre-edge structure in (Ti K-edge) X-ray absorption near edge structure (XANES) spectra of titanate nanotubes demonstrated that almost 40% of Ti atoms are uncoordinated rather than occupying symmetrical octahedral positions.[15] These uncoordinated centres are located on both surfaces of the nanotubes, as well as between the layers inside the nanotube walls.

3.1.2 Crystallography of Titanate Nanofibres, Nanorods and Nanosheets

During alkaline hydrothermal treatment of titanium dioxide the formation of fibrous, rather than tubular, nanostructures can occur. The XRD patterns of these nanofibrous materials usually show a characteristic reflection at small angles, confirming a layered structure of the nanofibre crystals. In current literature, the crystal structure of these nanofibres corresponds to the structure of layered protonated polytitanates $H_2Ti_nO_{2n+1}$, where $n = 5$ (ref. 16,17), 6 (ref. 18) and 8 (ref. 16,19), or else channelled TiO_2-B (ref. 20).

The nanofibres can be considered as a stack of titanate (100) planes. The length of the fibre usually corresponds to crystallographic axis c, which is considered as the direction of most rapid crystallization in polytitanates.[17] The width of the fibre usually corresponds to axes b and a. The interlayer distance in nanofibres is in the range of 0.7–0.8 nm. The calcination of titanate nanofibres results in the consecutive transformation from titanate to TiO_2-B (at 400 °C), then to anatase (at 700 °C) and to rutile (at 1000 °C). The nanofibrous morphology disappears at 1000 °C.[21]

Nanorods obtained by calcination of protonated titanate nanotubes at temperatures above 400 °C, are characterised by a tetragonal, anatase polycrystalline crystal structure with impurities of amorphous phase.

The crystal structure of nanosheets, which are observed as an intermediate product during nanotube or nanofibre synthesis, is attributed to either the hydrated form of delaminated anatase[22] or lepidocrocite-type titanates consisting of shared TiO_6 octahedrons.[23] In both cases, the simulated XRD patterns are similar to those of titanate nanotubes, making it difficult to discriminate between each particular structure. The use of the instrumental methods which can distinguish between three-coordinated oxygen atoms intrinsic to anatase structure, and four-coordinated oxygen atoms forming the characteristic sequence of edge-sharing TiO_6 octahedrons of lepidocrocite-type titanates, may enable an accurate determination of nanosheet crystal structure in the future.

3.1.3 Crystallography of Anodized and Template-Assisted TiO_2

The crystal structure of titanium oxide films produced by electrochemical oxidation of titanium can be amorphous or crystalline, depending on the parameters of preparation including: the applied potential, the time of ano-dising and the composition of the electrolyte. As far as TiO_2 nanotube arrays are concerned, as produced, they are usually characterised by an amorphous structure, which can be transformed to anatase at temperatures as low as 300 °C in air.[24,25] A further increase in temperature to above 450 °C, results in the formation of an anatase–rutile mixture.[25,26] Annealed nanotubes are char-acterised by a polycrystalline structure. The typical size of crystallites (deter-mined using TEM images or the Scherrer Equation for XRD data) varies in range from 10 to 25 nm. There is a degree of anisotropy in the nanotube array. The preferred orientation of the crystallite is such that plane (101) of anatase is found along the nanotube wall.[27]

The residual fluoride ions from electrolyte solutions are usually incorporated into amorphous TiO_2 nanotubes without formation of a fluoride phase. Annealing of nanotubes, however, leads to an almost complete loss of fluoride ions at 300 °C.[27]

Sol–gel template-assisted TiO_2 nanostructures are also usually characterised by an amorphous structure. Depending on the method of template removal, if pyrolysis is involved, the nanostructures under heat treatment may be con-verted to anatase or rutile.

3.1.4 Conclusions

Recently, more sophisticated crystallographic measurements, using total X-ray diffraction and the atomic pair distribution function obtained from synchrotron radiation,[28] have demonstrated that the three-dimensional structure of "TiO_2-type" nanotubular materials produced by the alkaline hydrothermal method can be interpreted as an arrangement of TiO_6 octahedra in corrugated layers. The particular arrangement of octahedrons may depend on the morphology of the nanostructures and the encapsulation of water or sodium ions. The results agree with the local structure of nanotubes obtained using EXAFS.[15] XANES studies have revealed that the assembly of TiO_6 octahedra is different from that in anatase, but some anatase-like structures may be presented in the nanotubes.[29]

It is likely that as-synthesised nanotubular materials correspond more closely to sodium titanates, than to anatase or TiO_2-(B). This conclusion is also sup-ported by (*i*) the frequent observation of a characteristic reflection at small angles in the XRD pattern;[18] (*ii*) the low isoelectric point (*ca.* 3) and the negative value of zeta potential in aqueous solutions, due to the acid–base dissociation of titanates;[30] (*iii*) the prefered adsorption of positively charged ions from aqueous solutions on the surface of nanotubes;[31] (*iv*) the very pro-nounced ion-exchange properties of nanotubes, allowing almost stoichiometric amounts of alkaline ions to be exchanged;[32] and (*v*) a dependence of the interlayer distance between the layers in the nanotube walls on the amount of alkaline ions contained within.[33,34]

The precise crystal structure of titanate nanotubes is still the subject of systematic investigations, including neutron diffraction studies. It is possible, however, to conclude that the structures have several common features. Firstly, it is a well defined, layered structure with a relatively large interlayer distance *ca.* 0.7–0.8 nm, resulting from the observation of a characteristic reflection (200) in the XRD patterns at small 2θ values of *ca.* 10°. Secondly, an atom of hydrogen situated in these interlayer cavities can be exchanged with alkali metal ions in the aqueous suspension. Thirdly, the layers of the (100) plane consist of edge- and corner-sharing TiO_6 octahedra, building up to zigzag structures. When protonated titanate nanotube samples are calcinated at moderate temperatures (less than 450 °C), the appearance of the TiO_2-B phase is likely. The phase transformation of titanate nanotubes and their stability will be discussed in Chapter 4.

3.2 Adsorption, Surface Area and Porosity

Due to the nanometre size of TiO_2 nanostructures, the proportion of surface atoms to bulk atoms is significant. Indeed, the typical dimensions of the unit cell of most TiO_2 compounds is approximately 0.3 nm. This means that in a nanometre size crystal, the line connecting two different points on the surface would go through the bulk of the crystal and would contain only several "volume" atoms. Such a dominance of the surface atoms is intrinsic to nanostructured materials.

3.2.1 Surface Area of Nanotubes

The specific surface area of some structures (*e.g.* nanotubular structures) can be estimated by the calculation of the geometrical surface area (see Figure 3.4). The specific surface area of a nanotube (S) can be calculated as a sum of the external (S_{ext}) and internal (S_{int}) surface areas divided by the mass of the nanotube using the formula:

$$S = \frac{S_{ext} + S_{int}}{\rho V_{tube}} \tag{3.1}$$

where ρ and V_{tube} are the density and the volume of the nanotube, respectively. For a cylindrical geometry, it is possible to express surface area and volume using the formulae:

$$S_{ext} + S_{int} = 2\pi r L + 2\pi (r + h)L = 2\pi (2r + h)L \tag{3.2}$$

and

$$V_{tube} = \pi (r + h)^2 L - \pi r^2 L = \pi h(2r + h)L \tag{3.3}$$

Where L and r are the the length and internal radius of the nanotube, respectively, and h is the thickness of the nanotube wall. Combining Equation

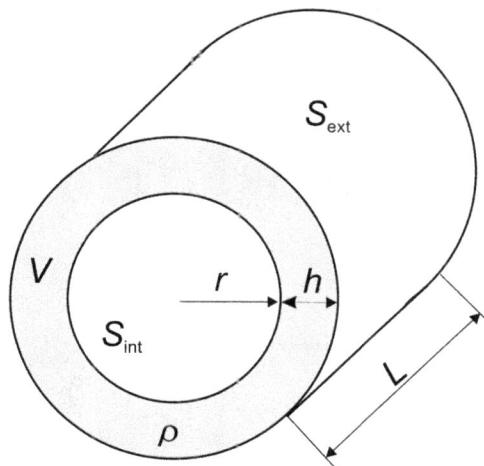

Figure 3.4 Diagram showing the geometrical characteristics of a nanotube for the calculation of specific surface area and pore volume: r is the radius of the nanotube, h is the wall thickness, L is the length of the nanotube, ρ is the density of the nanotubular material, V is the volume, and S_{ext} and S_{int} are the external and internal surface areas, respectively.

(3.1) with Equations (3.2) and (3.3), the specific surface area of the nanotube surface can be estimated as:

$$S = \frac{2}{\rho h} \qquad (3.4)$$

This value for the specific surface area of nanotubes does not depend on the internal or external diameter of nanotubes, but only on the wall thickness and density of the nanotubular materials. In practice, however, the nanotubes do not have an ideal cylindrical geometry and Equation (3.4) can only be used for guidance purposes.

The experimental surface area of materials can be determined using various adsorption methods, in which an adsorbate (probe species) with well characterised sorption properties is adsorbed on the surface of the material to be studied. By varying the adsorbate concentration (pressure) and temperature, and measuring the amount of probe material adsorbed, it is possible to determine the specific surface area, pore size distribution, heat or activation energy of adsorption, as well as the distribution of energy centres. One of the standard quantitative methods for the characterisation of a specific surface area is the BET (Brunauer–Emmett–Teller) surface area, obtained from the isotherm of nitrogen adsorption on the surface of porous materials at −195 °C.

A typical isotherm for nitrogen adsorption on the surface of titanate nanotube is shown in Figure 3.5. According to IUPAC recommendations,[35] the

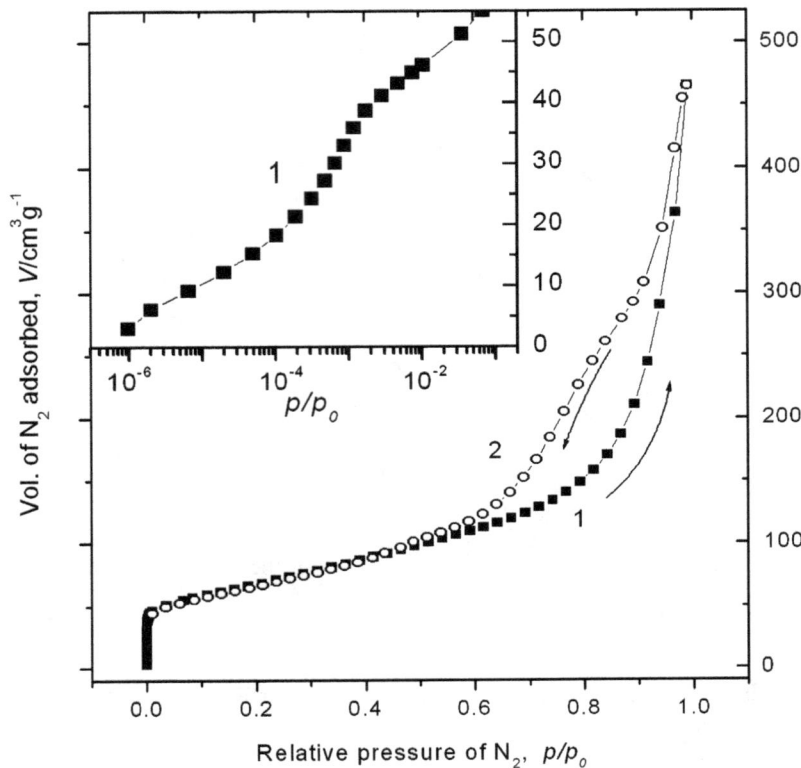

Figure 3.5 Isotherm of nitrogen adsorption (1) and desorption (2) at −196 °C on the surface of titanate nanotubes prepared by the alkaline hydrothermal method. The inset shows a semi-logarithmic plot of the adsorption isotherm. (Data are reproduced with kind permission from ref. 36).

type of hysteresis loop for the N_2 isotherms is intermediate between H1 (at $0.5 < p/p_0 < 0.8$) and H3 (at $p/p_0 > 0.8$). The H1 type is characteristic of uniform pores inside aggregates of particles. The observed hysteresis extended to $p/p_0 \approx 1$, indicates the presence of large pores which are not filled. Taking into account the morphology of the material observed by microscopy, the smaller pores may correspond to the pores inside the nanotubes, with the diameter of these pores being equal to the internal diameter of these nanotubes. The larger pores may correspond to the pores between the nanotubes. The hysteresis loop is relatively broad, indicating a wide distribution of pore sizes. This precludes speculation concerning the shape of the pores in the range of high relative pressures.

The BET surface area of titanate nanotubes determined from the nitrogen adsorption isotherm, typically varies between 200 and 300 $m^2 g^{-1}$, depending on the preparation method and the effectiveness of washing for the removal of sodium ions.[36] A similar range of specific surface areas can be obtained using

Equation (3.4), with a nanotube wall thickness, h, of 2 to 3 nm and a density, ρ (determined by a helium pycnometer), of $3.12 \, \text{g cm}^{-3}$. The density of titanate nanotubes can also be estimated from the parameters of a unit cell of $H_2Ti_3O_7$ ($a = 1.602$ nm, $b = 0.375$ nm, $c = 0.919$ nm and $\beta = 101.5°$, see Table 3.1), and taking into account that four "molecules" of trititanate occupy one unit cell. In this case, the value for density can be estimated as $3.16 \, \text{g cm}^{-3}$, which is close to that measured by helium adsorption.

For TiO$_2$ nanotube arrays obtained by the anodising method, it is difficult to obtain a BET surface area experimentally using nitrogen adsorption, since the impurities of metallic titanium distort the data. The value for the specific surface area, however, can be estimated from geometric considerations using Equation (3.4). For nanotubes having a typical wall thickness of 15 nm and the density of anatase being approximately $3.9 \, \text{g cm}^{-3}$, the value for the specific surface area can be estimated at *ca.* $35 \, \text{m}^2 \, \text{g}^{-1}$, which is approximately one order of magnitude smaller than the surface area of titanate nanotubes. The experimental value for the anodic TiO$_2$ surface area of nanotubes may be higher due to the porous structure of the walls.[37]

The specific surface area of titanate nanofibres produced by the alkaline hydrothermal treatment of TiO$_2$ at elevated temperatures, is usually smaller than that of titanate nanotubes and is *ca.* $20 \, \text{m}^2 \, \text{g}^{-1}$.[36]

3.2.2   Pore Volume of Nanotubes

Figure 3.6a shows the differential pore-volume distributions obtained from the desorption of nitrogen curve using the BJH algorithm, for two titanate nanotube samples characterised by different average tube diameters. The curves have a bell-like shape, and the maxima corresponds to pore sizes of 4 to 20 nm, relating this material to pores with a mesoporous range. In general, methods of nitrogen adsorption and, in particular, the BJH pore size distribution technique are suitable for characterising titanate nanotubes and determining their pore size distribution. A comparison of the pore size distribution of titanate nanotubes with two differing average diameters, obtained from nitrogen adsorption data with distribution built up from histograms using TEM images, is shown in Figure 3.6. In order to compare the distribution in the diameter of nanotubes with that of pore volume, the number of particles, N, was multiplied by the square of diameter, d^2, since the volume of nanotubes is proportional to d^2. It is apparent that the two distributions are slightly different, as seen in Figures 3.6a and 3.6b. The distribution obtained from the adsorption data yields larger nanotube diameter. However, the relative positions of the distributions for samples 1 and 2 are identical for both methods. Thus, it is possible to use BJH pore volume distribution for a semi-quantitative characterisation of the morphology of nanotubular titanates.

The disagreement between pore size distributions obtained from microscopy and adsorption data is attributable to the small number (60) of nanotubes included in the analysis of electron microscopy images. Only the nanotubes

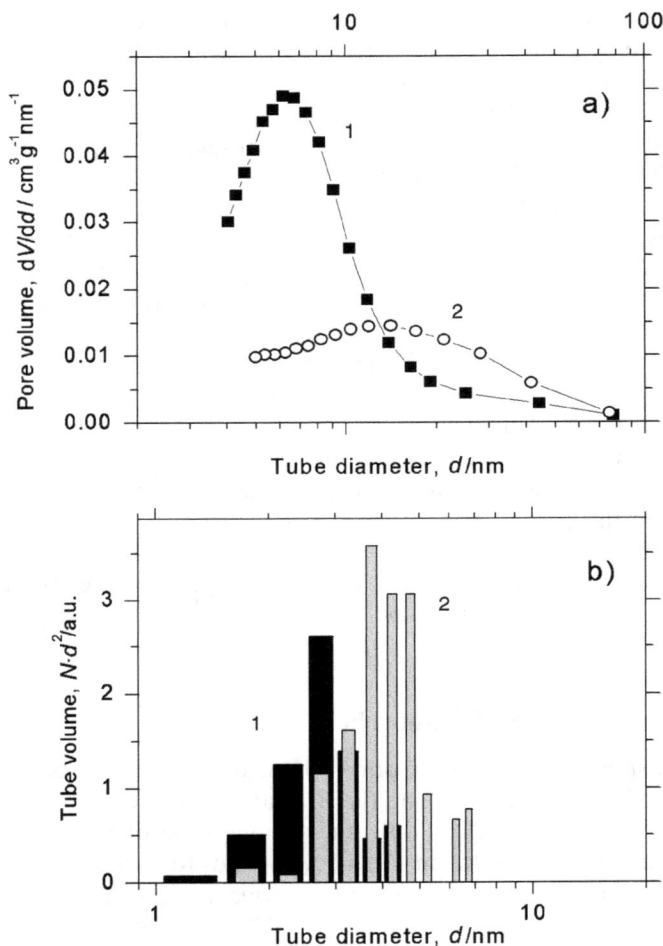

Figure 3.6 Comparison of a) pore volume distribution obtained from nitrogen desorption (data points) and b) nanotube internal diameter distribution obtained from TEM (histograms) for two samples of TiO_2 nanotubes produced by the hydrothermal treatment of (1) 0.25 g and (2) 3 g of anatase in 300 cm^3 of NaOH (10 mol dm^{-3}) at 140 °C. (Data are reproduced with kind permission from ref. 36).

which featured on a single image were considered, whereas the adsorption technique measures the entire sample. In this particular case, the larger nanotubes (with pore diameters of up to 20 nm) that were present in some samples did not feature in the histogram calculations, thus the mean tube diameter was underestimated. Furthermore, the BJH pore volume distribution includes not only the internal pores of nanotubes, but the larger pores formed between nanotubes, leading to an overestimation of mean tube volume. The BJH method is based on the principle that the surface tension of liquid nitrogen is constant and does not depend on the radius of meniscus (the Kelvin equation).

This may not, however, be true for pores of very small diameter (less than 4 nm). Hence, novel methods for determining pore size distribution should be applied, such as: a correction to the Kelvin equation,[38] a *t*-function[39] or DFT calculations.[40]

Distinction between internal and external pores

Pore volume distribution determined by the nitrogen adsorption method includes both internal (inside individual nanotubes) and external (cavities formed between nanotubes) pores. In order to distinguish each component, an ultrasonic treatment of samples of titanate nanotubes was used to break up the agglomerates into individual particles. From the SEM data shown in Figures 3.7c and 3.7d, it was found that two hours of ultrasonic treatment of an aqueous suspension of titanate nanotubes resulted in the destruction of the secondary

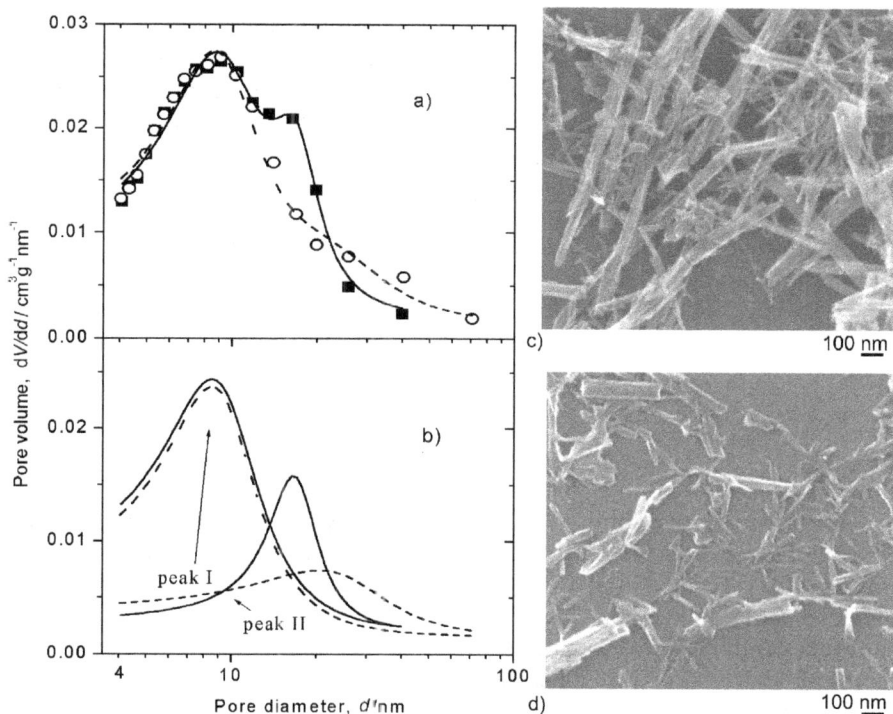

Figure 3.7   Pore-volume distribution (BJH desorption) of a) titanate nanotubes and b) its deconvolution into two Lorentzian curves. (○) initial sample, BET surface area of 199 m^2 g^{-1}, BJH desorption pore volume of $0.70 = 0.35$ (peak I) $+ 0.35$(peak II) cm^3 g^{-1}, (■) sample ultrasonically treated for 2 h, BET surface area of 198 m^2 g^{-1}, BJH desorption pore volume of $0.55 = 0.35$ (peak I) $+ 0.20$ (peak II) cm^3 g^{-1}. SEM images are shown for c) initial nanotubes and d) ultrasonically treated nanotubes. (Data are reproduced with kind permission from ref. 36).

structure of the material and a decrease in the average nanotube length. As a result, small nanotubes tended to assemble into close packing with very narrow pores between the tubes. A comparison of the pore volume distribution of both samples (before and after ultrasonic treatment) is shown in Figure 3.7. It is clear that the pore volume distribution has at least two components, and can be presented as a sum of two Lorentzian curves. The first component exhibits a maximum pore volume at a pore diameter of 8.5 nm and an integral of this component of $0.35\,cm^3\,g^{-1}$. It is proposed that this component corresponds to the internal pore volume distribution of the nanotubes, since it is not affected by the ultrasonic treatment.

The second component exhibits a maximum pore volume at a pore diameter of 20 nm and an integral of $0.35\,cm^3\,g^{-1}$ for the initial sample, and 16 nm and $0.20\,cm^3\,g^{-1}$ for the ultrasonically treated sample. It is likely that the second component corresponds to the pore volume distribution of pores formed between the nanotubes. Following the ultrasonic treatment, the agglomerates of titanate nanotubes were destroyed and some tubes were broken. This resulted in a separation of the individual nanotubes and the formation of a much more compact structure, with a smaller average distance between nanotubes compared with that in the initial sample. Consequently, the average pore diameter and total pore volume of external pores is also smaller. The ultrasonic treatment does not significantly change the BET surface area of the sample, since the increase in surface area due to an increasing of number of nanotube ends is negligible. In this fashion, it is possible to distinguish the contributions of internal pores and external pores to the total pore volume distribution function.

The specific volume of pores inside cylindrical nanotubes can be estimated, assuming a cylindrical nanotube geometry, using the following equation:

$$V_{int} = \frac{1}{\rho} \cdot \frac{r^2}{(r+h)^2 - r^2} \tag{3.5}$$

Taking the values of both $r = 3.5$ nm and $h = 2$ nm from TEM images, and a density of $\rho = 3.12\,g\,cm^{-3}$, the volume of the internal pores of titanate nanotubes can be estimated as $0.22\,cm^3\,g^{-1}$, which is less than the value $0.35\,cm^3\,g^{-1}$ obtained from Figure 3.7. This implies that the sample consists largely of nanotubes and the conversion of precursor anatase to titanate nanotubes during the alkali hydrothermal treatment is high. This conclusion is in agreement with the fact that no other phase was found on the TEM and SEM images.

3.2.3 Effect of Ionic Charge on Adsorption from Aqueous Solutions

In an aqueous suspension, nanotubular titanates tend to develop a negative zeta potential due to the dissociation of titanic acid (*e.g.* $H_2Ti_3O_7 = H^+ + HTi_3O_7^-$).[41] This phenomenon greatly affects the ability of charged molecules to adsorb onto the surface of nanotubes. The affinity of species in the cationic form is much better than that of anionic species, indicating the major role of electrostatic interactions between the adsorbent and the adsorbate during adsorption from aqueous solution.

Figure 3.8 shows adsorption isotherms for two organic dyes from aqueous solution onto the surface of titanate nanotubes and TiO₂ nanoparticles (P-25, Degussa). Methylene Blue (MB) is a cationic dye, whereas Eriochrome Black T (EBT) is an anionic one. The adsorption isotherms showed Langmuir-type adsorption, according to:

$$a' = \frac{a_s K_L C}{1 + K_L C} \tag{3.6}$$

where a_s corresponds to the amount of dye forming a monolayer on the surface of the adsorbent, and K_L is a constant of adsorption. When the concentration of MB in the solution exceeds $0.05 \, \mathrm{mmol \, dm^{-3}}$, the surface of titanate nanotubes is saturated with adsorbed MB molecules. The amount of adsorbed MB on the surface of nanotubes can be estimated as *ca.* $8 \, \mathrm{mmol(MB) \, mol(TiO_2)^{-1}}$.

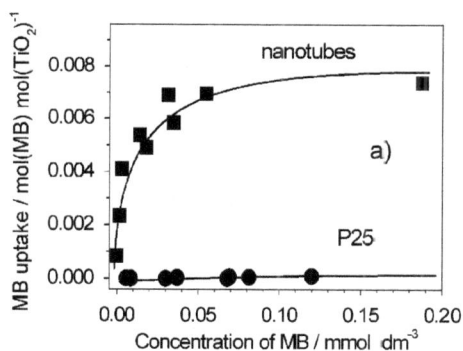

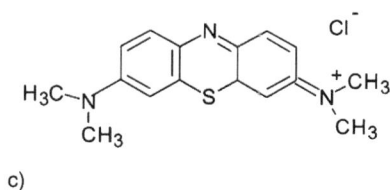

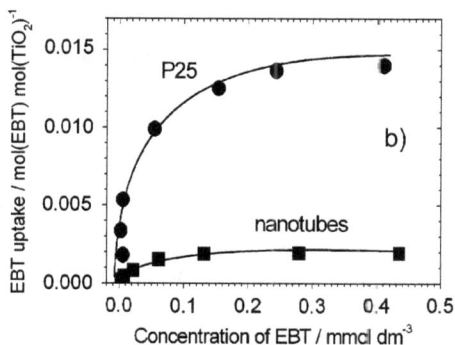

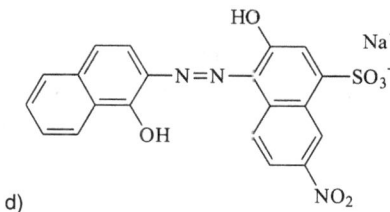

Figure 3.8 Isotherm for a) Methylene Blue (MB) and b) Eriochrome Black T (EBT) adsorption onto the surface of (■) titanate nanotubes and (●) spheroid P-25 TiO₂ nanoparticles from aqueous suspension at 25 °C. Structures are indicated for c) cationic MB and d) anionic EBT.

By contrast, the a_s value for MB on the surface of P-25 nanoparticles is almost two orders of magnitude smaller, at *ca.* $0.6\,\mu mol(MB)\,mol(TiO_2)^{-1}$. The values of the adsorption constant, K_L, and the saturation concentration, a_s, for nanotubes and nanoparticles are estimated by fitting data from Figure 3.8 a to Equation (3.6) and are shown in Table 3.2. Such a large difference in a_s values is difficult to explain by the difference in the values of specific surface area alone.

Figure 3.8b shows the isotherm of Eriochrome Black T (EBT) adsorption from an aqueous solution on the surface of nanotubes and nanoparticles. From Equation (3.6), the amounts of dye in monolayer a_s on the surface of nanotubes can be estimated (see Table 3.2). Despite the fact that the specific surface area of nanoparticles is almost a fifth that of nanotubes, the amount of EBT adsorbed onto the surface of the nanoparticles is greater than that adsorbed onto the nanotubes.

In aqueous solution, both dyes experience electrolytic dissociation. In the case of MB this leads to the formation of a cationic form of the dye over a wide pH range, whereas with EBT, an anionic form of the dye is produced (see Figure 3.8c and d). At the same time, titanate nanotubes and TiO_2 nanoparticles (P-25) suspended in aqueous media develop a zeta potential, which is negative for nanotubes[41] and positive for nanoparticles [the isoelectric point is 6.5 (ref. 42) in a pH range of 4 to 5]. Adsorption is preferred where the sign of the charge of dye and adsorbent is different, rather than where both have same sign. This indicates that electrostatic interactions dominate during the adsorption of charged dyes from aqueous media.

3.3 Electronic Structure of Titanate Nanotubes

The electronic structure of titanium dioxide has been thoroughly studied in recent years due to its potential use in several applications,[43] including: photocatalysis, photovoltaic cells, sensors and electronic devices. Generally, TiO_2 is a wide-band gap semiconductor ($E_G = 3.2\,eV$) with indirect interband electron transitions.[44] Titanium dioxide spheroidal nanoparticles show a relatively small apparent bandgap blue shift[45,46] (< 0.1–$0.2\,eV$), caused by quantum size effects for particles sizes down to 2 nm. Such small effects are mainly due to the relatively high effective mass of carriers in TiO_2 and an exciton radius in the approximate range of 0.75–1.90 nm,[47] such that only very small particles could possess an increased bandgap. It has also been suggested[48] that such a small blue shift of bandgap is due to the appearance of direct electronic transitions of small particles, rather than quantum confinement effects.

Recently,[49] a new form of TiO_2 exhibiting a two-dimensional structure with lepidocrocite TiO_2 flat nanosheets, was produced by exfoliation of layered protonic titanates.[50] The [51] electronic band structure of these nanosheets that was revealed was compared with the calculated[52] one. It was shown that the bandgap of nanosheets is strongly blue shifted ($E_G = 3.84\,eV$)[51] relative to the bandgap of bulk TiO_2, due to lower dimensionality, *i.e.*, a 3-D to 2-D transition.

Table 3.2 Adsorption properties of two organic dyes onto titanate nanotubes.

Dye	Adsorbent	$K_L{}^a/dm^3\,mmol^{-1}$	$a_s{}^a/mol(MB)\,mol(TiO_2)^{-1}$	BET specific area/$m^2\,g^{-1}$	Charges on dye/support
MB	Nanotubes	285	0.008	250	$+/-$
EBT	Nanotubes	31	0.002	250	$-/-$
MB	P-25	13	6×10^{-5}	50	$+/+$
EBT	P-25	132	0.014	50	$-/+$

^{a}Obtained using Equation (3.6)

The scrolling of two-dimensional nanosheets into one-dimensional nanotubes would naturally result in a dramatic change of the energy spectrum which, in turn, should result in a drastic modification of many of the electronic properties.[53] Indeed, when a 2-D sheet is rolled up to form a nanotube, the wave vector along the circumferential ($k_\perp$) direction becomes quantised, such that:

$$k_\perp = \frac{2n}{d} \qquad (3.7)$$

where d is the nanotube diameter and n is an integer. At the same time, the wave vector along the tube axis ($k_\parallel$) remains continuous. Due to this quantisation, the energy bands that are produced by the surfaces in the k-space in a 2-D case, are reduced to a set of sub-bands. This is illustrated in Figure 3.9a, where the valence and conduction bands of a 2-D semiconductor are schematically represented by two paraboloids. The energy spectrum of nanotubes in this case is a set of parabolas (as shown in Figure 3.9b), which are formed from cross-sections of the 2-D bands by a set of parallel planes, each of which corresponds to a different value of the quantised vector $k_\perp$ and is oriented perpendicular to the $k_\perp$–vector. The separation of the subbands along both the $k_\perp$ and energy axes depends on the nanotube diameter, increasing when d decreases.

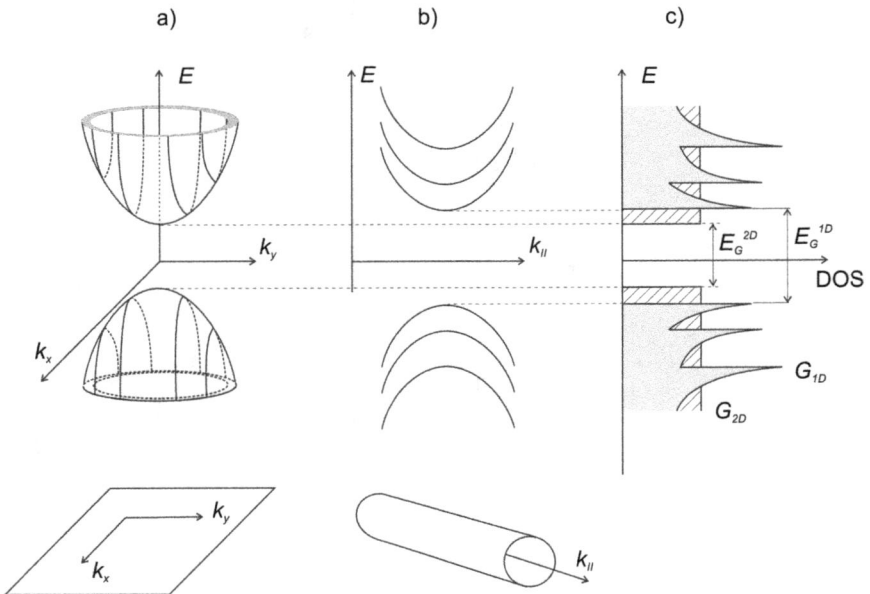

Figure 3.9 The transformation of the electron band structure of nanosheet semiconductors accompanying the formation of nanotubes. Band diagrams of a) a 2-dimensional nanosheet and b) quasi-1-dimensional nanotubes are shown. Diagram c) shows the energy density of states for nanosheets (G_{2D}) and nanotubes (G_{1D}). E_G^{1D} and E_G^{2D} are the band gaps of 1-D and 2-D structures, respectively (see Equation 3.11), and k_x and k_y are the wave vectors.

Within the effective mass model, the energy spectrum of 2-D TiO$_2$ sheets can be described by:[54]

$$E_{2D}^{\pm} = \pm \frac{E_G}{2} \pm \frac{\hbar_P^2 k^2}{2m_{e,h}} \tag{3.8}$$

where the 'plus' and 'minus' signs correspond to the conduction and valence bands, respectively, E_G is the energy gap, $\hbar_P$ is the reduced Planck's constant; m_e and m_h are the effective masses of electrons and holes, respectively. The electronic band structure of a TiO$_2$ nanotube, can be obtained from this relation by zone-folding and is given by a series of quasi 1-D sub-bands with different indices, n, as seen in Figure 3.9b:

$$E_{n1D}^{\pm}(k) = \pm \frac{E_G}{2} \pm \frac{\hbar_P^2}{2m_{e,h}} \left[k_{//}^2 + \left(\frac{2n}{d} \right)^2 \right] \tag{3.9}$$

This transition from a 2-D to a quasi 1-D energy spectrum has a dramatic effect on the energy density of states. In the 2-D case, the density of states, $G_{2D} = m_{e,h}/\pi\hbar_P^2$, has a constant value[54,55] for energies outside the energy gap, as indicated in Figure 3.9c. In the quasi 1-D case, however, the density of states in each sub-band is given by:

$$G_{n,1D}(E) = \pm \left\{ \frac{m_{e,h}}{2\pi^2 \hbar_P^2 [E - E_n(0)]} \right\}^{1/2} \tag{3.10}$$

This diverges at the band edge $E_n(0)$ leading to van Hove singularities.[54] The resulting density of state is formed by a series of sharp peaks with long overlapping tails, as shown in Figure 3.9c. The energy gap between the valence and conductance bands in the quasi-1-D case is larger than that in the parental 2-D material, and this difference increases with decreasing diameter of the nanotube.

The change in the energy gaps after rolling of a nanosheet to a nanotube can be expressed as:

$$\Delta E_G = E_G^{1D} - E_G^{2D} = \frac{2\hbar_P^2}{d^2} \left(\frac{1}{m_e} + \frac{1}{m_h} \right) \tag{3.11}$$

3.3.1 Spectroscopy of Titanate Nanotubes: UV/VIS, Pl, ESR, XPS, NMR, Raman and FTIR

The electronic structure of titanate nanotubes has been studied using a variety of different experimental techniques. Figure 3.10 shows the optical absorption spectra of a colloidal solution of titanate nanotubes of different average internal diameters.[53] The absorption spectra are very broad with several features at 250 and 285 nm. The threshold of the spectra suggests that the

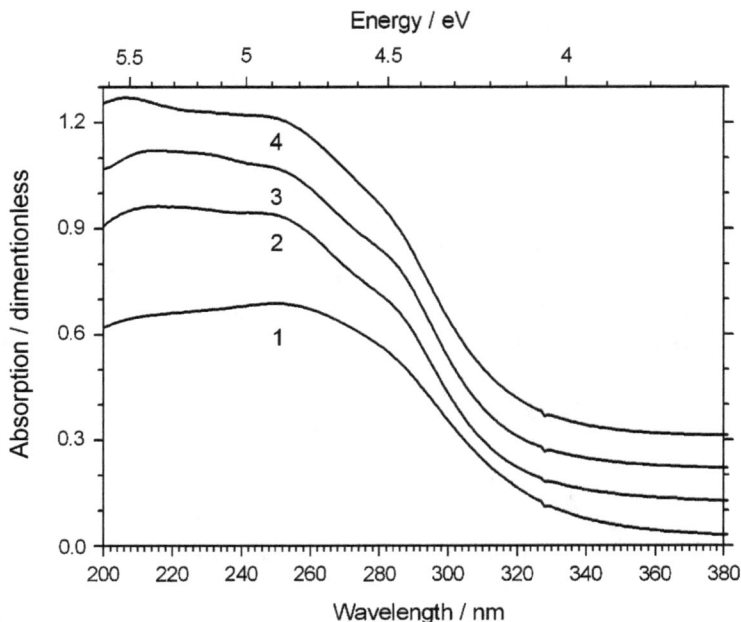

Figure 3.10 Visible absorption spectra of colloidal titanate nanotubes of differing mean internal diameter: 1) 2.5 nm, 2) 3.1 nm, 3) 3.5 nm, and 4) 5 nm. Path length: 1 cm and temperature: 25 °C. The curves are shifted vertically for clarity. (Data are adapted from ref. 53).

estimated value for the band gap of titanate nanotubes is approximately 3.87 eV. The position of the two features in the spectra and the apparent band gap do not change for nanotube samples of different internal diameter, indicating the absence of systematic size-dependent changes. These absorption spectra are also similar to the spectra of titanate nanotubes suspended in absolute ethanol solution,[56] but contrast with the diffuse reflectance spectra obtained from solid films, in which the effect of elastic light scattering can cause additional error.

Usually an increase in the average internal diameter of titanate nanotubes results in an increase in the average thickness of the multilayered nanotube wall, which is proportional to the number of layers in the wall. Recently,[93] it has been observed that an increase in the number of layers in multilayered lepidocrocite titanate films (produced by the deposition of single layer nanosheets) does not change the electronic structure of this film, which is similar to the case of single layer nanosheets. The stacking of several nanosheets does not affect the electronic properties of the solid, probably as a result of the weak electronic interaction between layers. A similar situation of weak interaction between different layers is likely to exist in the case of multilayered titanate nanotubes.

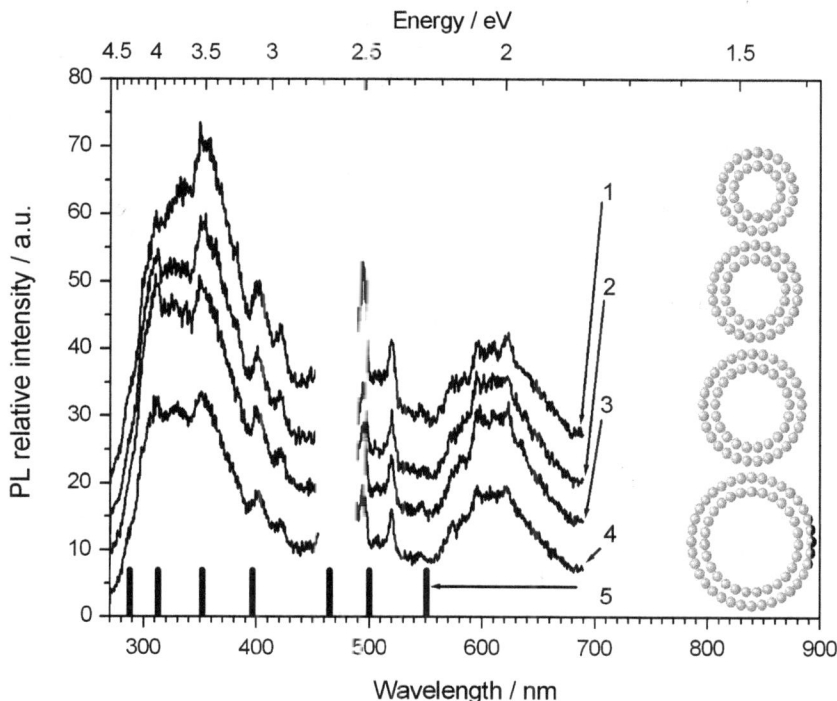

Figure 3.11   Photoluminescence spectra of colloidal titanate nanotubes of differing mean internal diameter: 1) 2.5 nm, 2) 3.1 nm, 3) 3.5 nm, and 4) 5 nm. Temperature: 22 °C, excitation wavelength: 237 nm and slit width: 5 nm; the wavelength range of 455–490 nm is omitted due to the high signal of the second harmonic from scattered excitation light. The curves are shifted vertically for clarity. 5) Vertical lines show the position of peaks in PL spectrum of titanate nanosheets. (PL spectra are adapted from ref. 53 for nanotubes and ref. 49 for nanosheets).

Figure 3.11 shows the photoluminescence spectra of aqueous colloids of titanate nanotubes with average internal diameters varying from 2.5 nm to 5 nm.[53] The spectra show several sharp bands across a wide range of wavelengths. The positions and relative intensity of all of these bands are practically identical and do not depend on the average internal diameter of nanotubular titanates. The shape and position of some peaks in the luminescence spectrum are similar to the peaks in the luminescence spectrum of the colloidal solution of exfoliated lepidocrocite titanate nanosheets,[49] see Figure 3.11.(5). The best match is exhibited for peaks at shorter wavelength (300–400 nm). At longer wavelengths (500–600 nm), the luminescence peaks for the nanotubes differ from those of the nanosheets. This could be attributable to the nature of these low energy states as cations can be incorporated between two or more nanosheets[57] and the spectra consequently depend on the composition of surrounding media. Indeed, titanate nanotubes produced by hydrothermal alkali treatment contain a residual amount of sodium ions,[58] whereas nanosheets produced by the exfoliation of layered titanates can be contaminated with caesium and tetrabutylammonium ions.[50]Since luminescence from low energy states is very sensitive to the presence of cations in the solution, the differences in the luminescence spectrum between nanotubes and nanosheets at longer wavelengths may be due to cation incorporation.

The independence of bandgap and energy levels positions on the diameter of nanotubes, as well as the spectral similarities between nanotube and nanosheets, suggest that all values for energy differences, ΔE_G, in Equation (3.11) are relatively small when compared with the accuracy of experiments. Indeed, the effective mass of charge carriers in bulk TiO_2 crystals is relatively large. According to different sources, the effective masses of electrons, m_e, can vary between $5m_0$ (ref. 59) and $30m_0$ (ref. 60), and the mass of holes, m_h, is greater than $3m_0$ (ref. 48,61). Taking $m_e = 9m_0$ and $m_h = 3m_0$ and using Equation (3.11), a difference of 8 meV in the energy gap of nanotubes of diameter 2.5 and 5 nm can be obtained. The energy difference between the two first peaks in the density of states $G_{1D}(E)$ (see Figure 3.9) is <24 meV for $d = 2.5$ nm, and 6 meV for $d = 5$ nm. The values obtained are too small to be resolved in experiments at ambient room temperature. Both the shift of band gaps and the $G_{1D}(E)$ peaks should be completely smeared by thermal fluctuations that have an amplitude $kT = 26$ meV.

The oxidation state of titanium in titanate nanotubes is predominantly equal to $4+$, however some impurities of Ti^{3+} ions in the lattice of titanate can be formed under calcination in vacuum. Figure 3.12 shows the electron spin resonance (ESR) spectra[62] of protonated and sodium titanate nanotubes annealed at elevated temperatures in a vacuum. The titanate nanotubes produced do not usually have paramagnetic signals. However, calcination of the sample in air may result in the appearance of a symmetrical peak characterised by $g = 2.003$ in the ESR spectrum, which can be attributed to the signal from single-electron-trapped oxygen vacancies (SETOV).[63] The maximum intensity of the SETOV signal is usually detected at 450 °C, when the transformation of titanate to anatase occurs (see Chapter 4 for information about the temperature stability of nanotubes). Calcination of H-TiNT in a vacuum results in the appearance of symmetrical ($g = 2.003$) and asymmetrical ($g = 1.98$) signals in the ESR spectrum (see Figure 3.12a; ref. 62). The latter is usually associated with the reduction of Ti^{4+} to Ti^{3+} ions on the surface of TiO_2, accompanied by the removal of oxygen atoms from lattice, [64] which is consistent with the disappearance of this signal after a long exposure of samples in an air atmosphere. In contrast, calcination of sodium titanate nanotubes results in the appearance of the SETOV signal only (see Figure 3.12b), due to the possible improved stability of sodium saturated nanotubes towards the removal of oxygen atoms.

The small impurities of Ti^{3+} ions in titanate nanotubes have also been detected using X-ray photoelectron spectroscopy (XPS). Figure 3.13a shows a typical XPS spectrum of titanate nanotubes annealed at 500 °C in a N_2 atmosphere.[65] The spectrum shows the region of the Ti 2p signals. The peak at

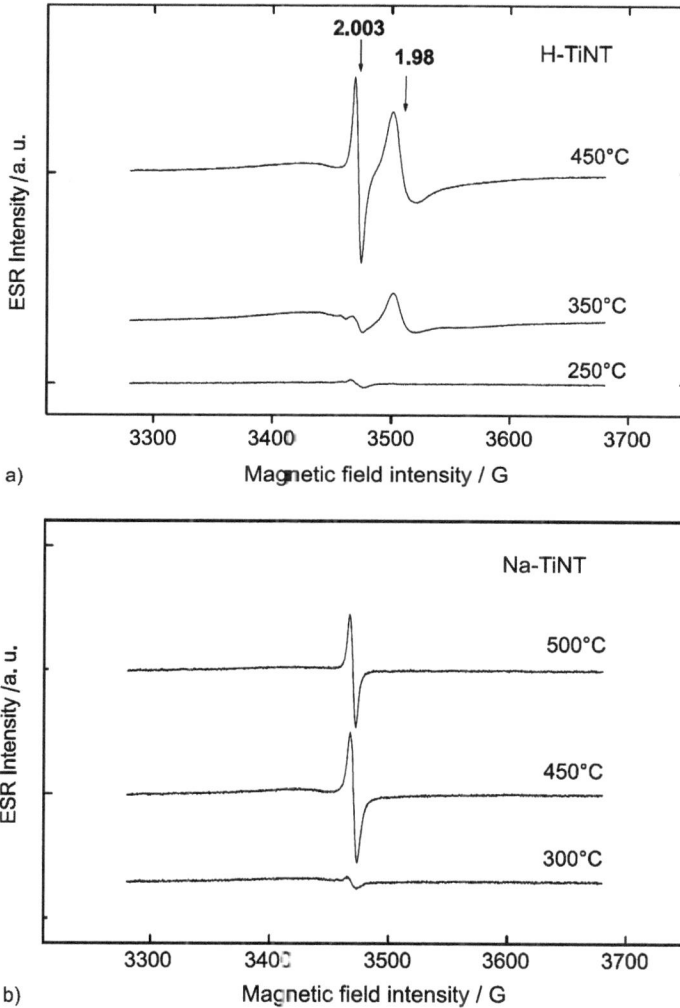

Figure 3.12   ESR spectra of calcined titanate nanotubes under vacuum conditions: a) protonated titanate nanotubes and b) sodium titanate nanotubes. (Spectra are adapted from ref. 62).

a binding energy of 459.5 eV can be attributed to the Ti^{+4} 2p$_{3/2}$ state. The reduction of Ti^{4+} to Ti^{3+} ions results in the shift of this peak to lower energies.[66] In Figure 3.13a, the peak at 459.1–459.4 eV is associated with Ti^{4+} cations, while the small peak at 457.5–458.1 eV corresponds to reduced Ti^{3+} ions. The ratio of Ti^{3+} to Ti^{4+} ions increases from 0.026 at 110 °C to 0.06 at 500 °C,[65] this suggests the removal of oxygen atoms from the lattice and the partial reduction of titanate nanotubes under anaerobic conditions at elevated temperatures. Surprisingly, even uncalcined samples have Ti^{3+} ion impurities.

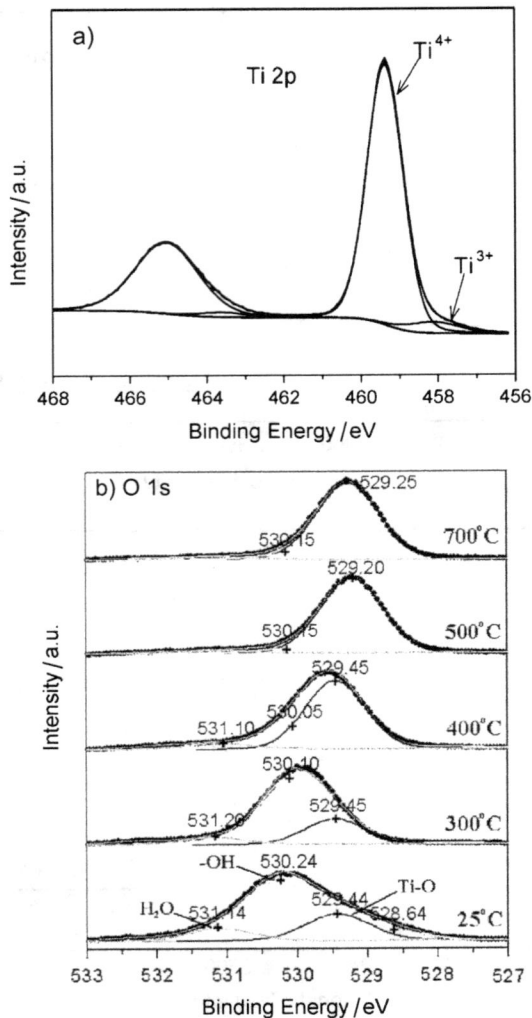

Figure 3.13 Photoelectron spectra of titanate nanotubes: a) calcined at 500 °C under
a nitrogen atmosphere, Ti 2p range, and b) annealed at various temperatures in air, O 1s range. (Data are adapted from refs. 65 and 67).

XPS spectroscopy also helps to reveal the nature of the oxygen states in
titanate nanotubes. Figure 3.13b shows O 1s signals in the XPS spectra of
titanate nanotubes annealed at different temperatures under an air atmosphere.[67,68] This XPS signal can be deconvolved into four subpeaks at at
531.14, 530.24, 529.44 and 528.64 eV, which can be attributed to H_2O, –OH,
Ti–O and an unidentified peak, respectively.[67] When the tubes are annealed at
300 °C, the intensity of the H_2O peak is reduced due to evaporation of both
physically adsorbed and crystallographic water. The unidentified peak also

disappears. The intensity of the –OH bonding peak decreases, with the increase of the Ti–O peak intensity when annealed at 400 °C, probably due to the dehydration of interlayer OH groups and thus phase conversion to the nanorods of the anatase phase. Above 500 °C, only the main peak assigned to Ti–O at 529.2 eV is observed.

The three different forms of oxygen atoms (H$_2$O, –OH and Ti–O) in protonated titanate also suggest that there are several forms of hydrogen atoms in the structure of nanotubes. Indeed, taking into account crystallographic water, the general structure of titanate nanotubes can be presented as H$_{2m}$Ti$_n$O$_{2n+m}$·xH$_2$O, where there are at least two type of protons existing in the structure, namely: ion-exchangeable acidic protons and the protons of crystallographic water. Our recent H$^-$-MAS NMR studies of sodium ion-exchanged titanate nanotubes have enabled us to identify the different forms of hydrogen in the spectrum.[69] Figure 3.14 shows the NMR spectra of protonated titanate nanotubes with a variable amount of ion-exchanged sodium ions. An increase in the sodium content of titanate nanotubes was found to result in a decrease in peak intensity at 10 ppm, accompanied by a simultaneous increase in intensity and shift of the peak at 6.3 ppm to 5.1 ppm (see Figure 3.14). Such behaviour suggests that ion-exchangeable protons have a resonance in a weaker magnetic field at 10 ppm and the chemical shift of these protons is unaffected by

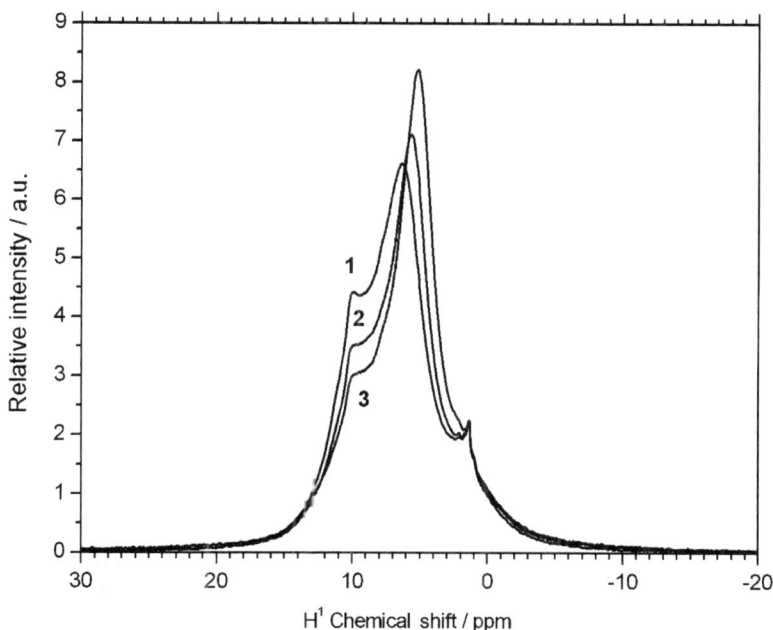

Figure 3.14   H^1 MAS NMR spectra of protonated titanate nanotubes which have been ion-exchanged with NaOH at 1) pH = 2, 2) pH = 9 and 3) pH = 13, followed by drying at 140 °C. (Several peaks in the chemical shift region of 1.3 ppm are signals from the NMR rotor).

the presence of sodium ions. In contrast, the addition of sodium ions results in an increase in the amount of crystallographic water, which is consistent with recent XRD studies,[70] as well as a change in the local magnetic field shielding, due in all probability to the close distance between the protons in crystallographic water and the exchanged sodium ions.

The abundance of OH groups in titanate nanotubes can also be confirmed by infrared spectroscopy measurements. Figure 3.15 shows typical FTIR spectrum of protonated titanate nanotubes.[71,72] The presence of crystallographic water molecules in the sample is confirmed by the appearance of a characteristic peak at 1630 cm^{-1}, that can be assigned to the H–O–H deformation mode (δ_{H-O-H}).

The broad intense bands at 3418 cm^{-1} and 3169 cm^{-1} can be attributed to surface OH stretching mode oscillations. As a result of the strong interaction between Ti ions and OH groups within the tubular structure, a shoulder at 3169 cm^{-1} from Ti–OH bonds is observed.[71]

The profile of these two bands also suggests a complex hydrogen bond environment in titanate nanotubes. The wide band at 501 cm^{-1} can be assigned to the crystal lattice vibration of TiO$_6$ octahedra, which can also be strongly affected by the incorporation of ions (*via* ion-exchange) into the titanate nanotubes,[32,72] or by the alteration of nanotubular morphology, including the diameter of the nanotubes.

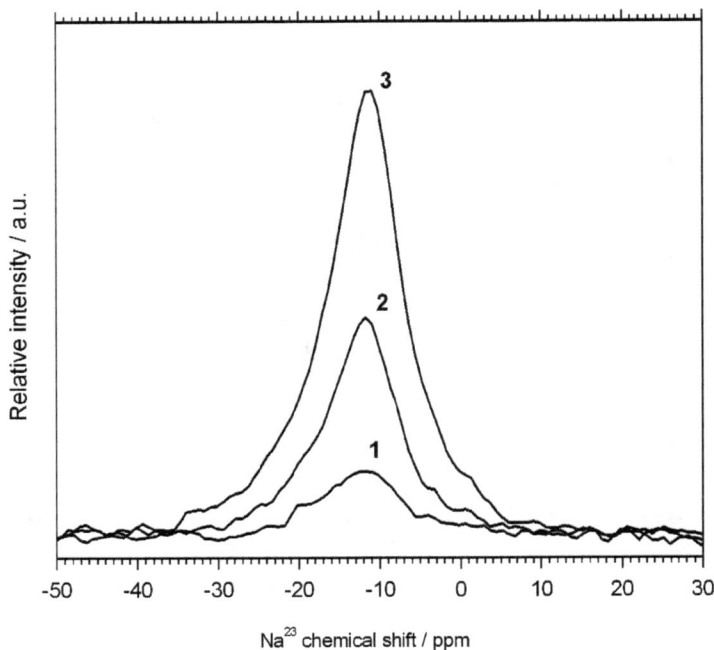

Figure 3.15 Na^{23} MAS NMR spectra of protonated titanate nanotubes which have been ion-exchanged with NaOH at: 1) pH = 2, 2) pH = 9 and 3) pH = 13.

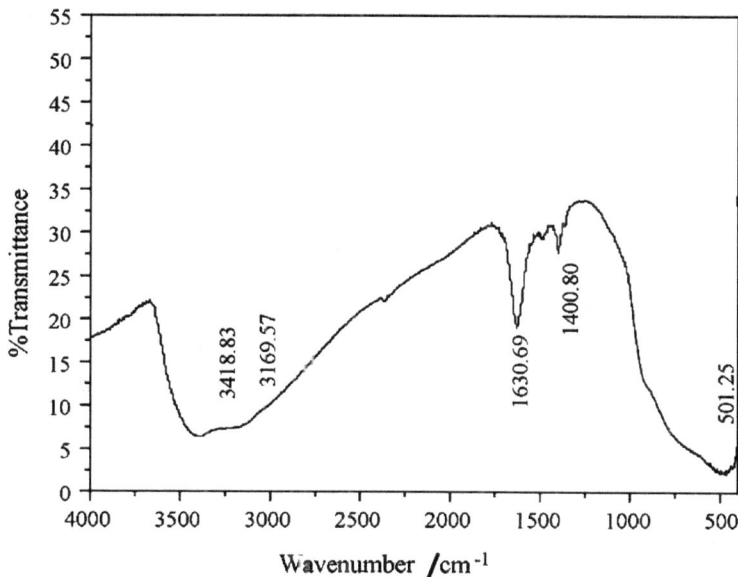

Figure 3.16 An FTIR spectrum of protonated titanate nanotubes. (Spectra are adapted from ref. 71).

The increase of sodium content in titanate nanotubes does not affect the position and the width of the resonance signal from Na[23] nucleuses (see Figure 3.16) suggesting a small interaction between the ions due to the large distance. The chemical shift of the single peak is *ca.* −12 ppm and the peak full width at half maximum height (FWHM) is approximately 10.7 ppm.

The typical[73] Raman spectrum of titanate nanotubes (see Figure 3.17) has two sharp, high intensity peaks at 290 cm^{-1} and 448 cm^{-1}, one multiplet peak centred at 668 cm^{-1} and several peaks of lower intensity at 388 cm^{-1}, 827 cm^{-1} and 917 cm^{-1}. This spectrum is similar to the recently observed spectrum of titanate nanotubes[22,74,75] and cannot be superimposed onto the Raman spectra of anatase and rutile. The exact assignment of these peaks is still under discussion, but they could be interpreted as Ti–O–Ti crystal phonons (most likely for the peaks at 448 cm^{-1} and 668 cm^{-1}) or Ti–O–Na vibrations (most likely for the peak at 917 cm^{-1}).[65] The second-order harmonics or radial "breathing" oscillations intrinsic to nanotubular structures may also be the source of these peaks.[76]

3.3.2 Electrical, Proton and Thermal Conductivities of Titanate Nanotubes

The electrical conductivity of carbon nanotubes strongly depends on chirality and can vary over several orders of magnitude. As a consequence, the properties of nanotubes can change from those of a conductive metal to those of a

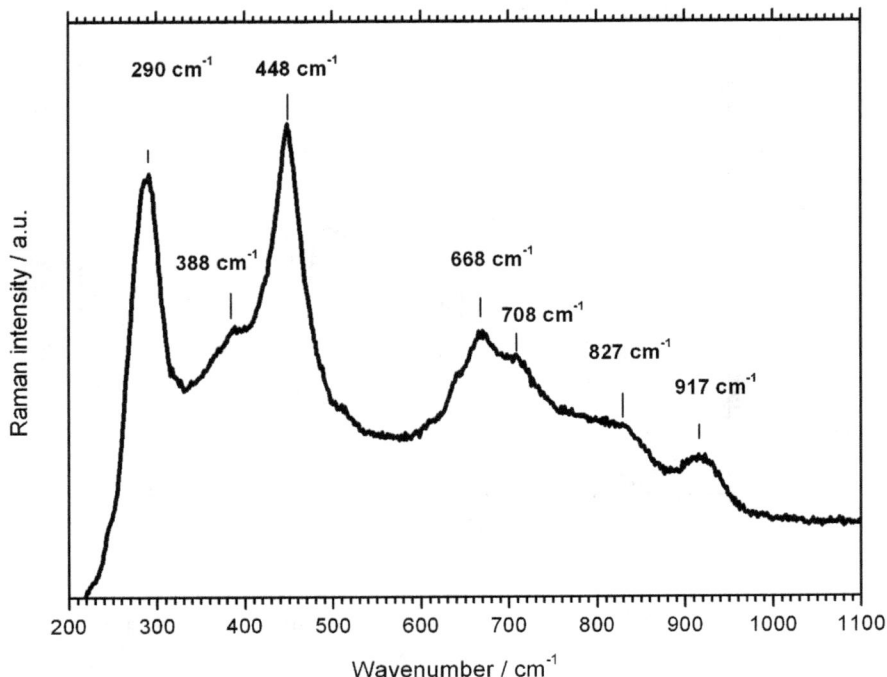

Figure 3.17   A typical Raman spectrum of titanate nanotubes. (Spectra are adapted from ref. 73).

semiconductor. For titanate nanotubes, the existence of labile ion-exchangeable protons suggests that this material should have pronounced proton conductivity. Using AC impedance spectroscopy, it was demonstrated[77] that proton conductivity dominated at temperatures below 130 °C in air as a pressed powder pellet of titanate nanotubes. The value of conductivity was *ca.* 5.5×10^{-6} S cm^{-1} at 30 °C, and increased with temperature to 1.5×10^{-5} S cm^{-1} at 130 °C. A further increase in the temperature resulted in removal of physiadsorbed water from the pores of nanotubes, which coincides with a dramatic fall in conductivity to a value of approximately 5.6×10^{-8} S cm^{-1}. This low conductivity represents the electron conductivity of titanate nanotubes. The conductivity increased at higher temperatures, having an apparent activation energy of 0.57 eV and is higher than the conductivity of anatase or rutile nanoparticles. For instance, at a temperature of 225 °C, the conductivity for titanate nanotubes was *ca.* 7.9×10^{-7} S cm^{-1} (ref. 77), whereas the value for anatase nanoparticles of 6 nm diameter was only *ca.* 10^{-9} S cm^{-1} (ref. 78).

Despite titanate nanotubes having a higher electrical conductivity than TiO_2, their conductivity can still limit their application in fields where a higher electrical conductivity is required (see Chapter 6 for details). There are, however, several approaches to increase the electroconductivity of nanotubes. A reduction of titanate nanotubes at an elevated temperature under vacuum, or in a hydrogen atmosphere, results in the slow removal of lattice oxygen and the formation of oxygen vacancies. This process is usually accompanied by an increase in absorption in the visible range of the absorption spectrum, due to the formation of Ti^{3+} centres, which can form Magneli phases (mainly Ti$_4$O$_7$ or Ti$_5$O$_9$). From a practical point of view, the synthesis of nanotubes containing significant Magnéli phases would be of great interest since it would dramatically increase their electrical conductivity.[79] The reverse electrochemical reduction of Ti^{4+} to Ti^{3+}, accompanied by a change in colour of the transparent nanotubular titanate film deposited on electroconductive glass can be exploited in electrochromic devices.[80] Another method for improving electrical conductivity is to prepare composites with other conductive materials, including: metals, carbon or conductive polymers.

By analogy with carbon nanotubes, one would expect to observe interesting thermoconductive properties of low-dimensional titanate nanotubes. Preliminary studies of the specific heat of titanates nanotubes have demonstrated that, at temperatures below 50 K, low-dimensional behaviour of acoustic phonons dominates, resulting in a large enhancement of their specific heat compared to bulk anatase or rutile.[81] An aqueous suspension of titanate nanotubes (2.5 wt% of nanotubes) also demonstrates a small thermal conductivity enhancement of 3% at 25 °C and 5% at 40 °C.[82]

3.4   Physical Properties of TiO$_2$ Nanotube Arrays

Physico-chemical properties of TiO$_2$ nanotube arrays are very close to the properties of TiO$_2$ nanoparticles, due to chemical and structural similarities. However, some unusual behaviour of nanotube arrays is observed, resulting from the specific nanotubular morphology of materials, and is considered in this section.

Figure 3.18 shows the typical transmittance spectra of a 400 nm thick TiO$_2$ nanotube array film on glass substrate.[83] The optical behaviour of the TiO$_2$ nanotube array is quite similar to that of mesostructured titanium dioxide. The TiO$_2$ nanotube absorption coefficient, α', can be calculated from transmittance, T':

$$\alpha' = -2.303 \frac{log(T')}{L} \tag{3.12}$$

where L is the film thickness (or the length of the nanotubes). The absorption coefficient depends on the energy of the incident light ($h_P\nu$) as follows:

$$\alpha' = A' \frac{(h_P\nu - E_G)^n}{h_P\nu} \tag{3.13}$$

where E_G is the semiconductor bandgap, A' is a proportionality coefficient, n is a number (which is equal to 1/2, 2, 3/2 or 3 for allowed direct, allowed indirect, forbidden direct or forbidden indirect transitions in the semiconductor,

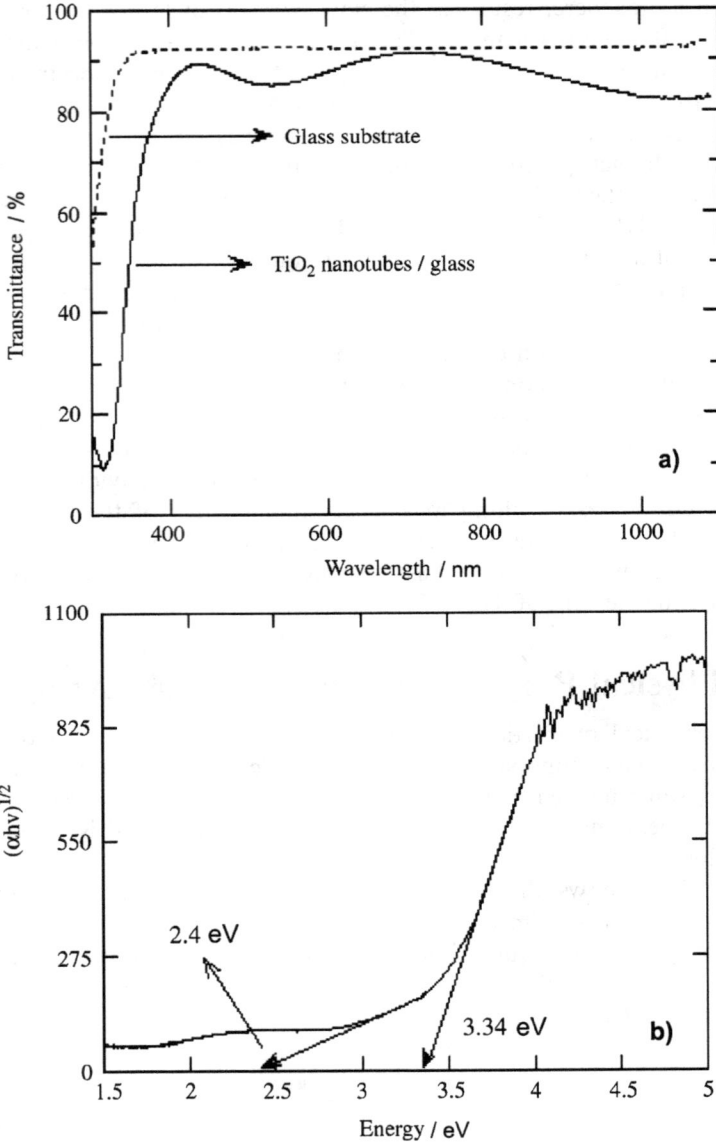

Figure 3.18   Optical properties of TiO$_2$ nanotube arrays: a) a transmittance spectrum of a 400 nm thick film on a glass substrate, b) the Tauc plot. (Data are adapted from ref 83).

respectively). TiO$_2$ nanotubes have allowed indirect optical transitions and the Tauc plot in $(\alpha h_P v)^{0.5}$ vs. hv coordinates allow us to estimate the optical band gap by dropping a line from the maximum slope of the curve to the x-axis, as 3.34 eV as seen in Figure 3.18b. The reported band gap value of anatase phase in bulk is 3.2 eV (ref. 44). A slight blue shift in the value might be due to a quantization effect in the nanotubular film where the wall thickness is about 12 nm. A band tail to 2.4 eV is observed. The degree of lattice distortion is likely to be relatively higher for nanotube array films, thus causing an aggregation of vacancies acting as trap states along the seams of nanotube walls leading to a lower band-to-band transition energy.

The uniformity of the TiO$_2$ nanotube array film thickness may also result in appearance of an interference pattern in the absorption spectrum of nanotubes (see the features at 475 and 1100 nm in Figure 3.18a), where the wavelength of the incident light and the thickness of the film have values in the same order magnitude. The spacing between the interference patterns, which is created by the interaction of the transmitted wave and the wave reflected back from the top of the nanotubes, reduces with increasing nanotube length. The position of the interference peaks also affected the thickness of the film.

The electrical conductivity of the TiO$_2$ nanotube array is comparable with that of conventional mesoporous TiO$_2$ films. However, the conductivity can increase by several orders of magnitude in the presence of hydrogen at room temperature, which interacts with the surface of the nanotubes and modifes their properties. In Chapter 6, details of such hydrogen sensing materials are provided.

References

1. T. Kasuga, M. Hiramatsu, A. Hoson, T. Sekino and K. Niihara, *Langmuir*, 1998, **14**, 3160.
2. B. Poudel, W. Z. Wang, C. Dames, J. Y. Huang, S. Kunwar, D. Z. Wang, D. Banerjee, G. Chen and Z. F. Ren, *Nanotechnology*, 2005, **16**, 1935.
3. G. H. Du, Q. Chen, R. C. Che, Z. Y. Yuan and L. -M. Peng, *Appl. Phys. Lett.*, 2001, **79**, 3702.
4. Q. Chen, G. H. Du, S. Zhang and L. M. Peng, *Acta Crystallogr., Sect B: Struct. Sci.*, 2002, **58**, 587.
5. D. Wu, J. Liu, X. Zhao, A. Li, Y. Chen and N. Ming, *Chem. Mater.*, 2006, **18**, 547.
6. A. Nakahira, W. Kato, M. Tamai, T. Isshiki, K. Nishio and H. Aritani, *J. Mater. Sci.*, 2004, **39**, 4239.
7. J. J. Yang, Z. S. Jin, X. D. Wang, W. Li, J. W. Zhang, S. L. Zhang, X. Y. Guo and Z. J. Zhang, *Dalton Trans.*, 2003, **20**, 3898.
8. R. Z. Ma, Y. Bando and T. Sasaki, *Chem. Phys. Lett.*, 2003, **380**, 577.
9. R. Z. Ma, K. Fukuda, T. Sasaki, M. Osada and Y. Bando, *J. Phys. Chem. B*, 2005, **109**, 6210.
10. Y. Kubota, H. Kurata and S. Isoda, *Mol. Cryst. Liq. Cryst.*, 2006, **445**, 107.
11. Q. Chen and L. M. Peng, *Int. J. Nanotechnol.*, 2007, **4**, 44.
12. R. Marchand, L. Brohan and M. Tournoux, *Mater. Res. Bull.*, 1980, **15**, 1129.

13. G. Armstrong, A. R. Armstrong, J. Canales and P. G. Bruce, *Chem. Commun.*, 2005, **19**, 2454.
14. M. Zukalova, M. Kalbac, L. Kavan, I. Exnar and M. Graetzel, *Chem. Mater.*, 2005, **17**, 1248.
15. Z. V. Saponjic, N. M. Dimitrijevic, D. M. Tiede, A. J. Goshe, X. Zuo, L. X. Chen, A. S. Barnard, P. Zapol, L. Curtiss and T. Rajh, *Adv. Mater.*, 2005, **17**(8), 965.
16. Z. Y. Yuan and B. L. Su, *Colloids Surf., A*, 2004, **241**, 173.
17. H. G. Yang and H. C. Zeng, *J. Am. Chem. Soc.*, 2005, **127**, 270.
18. X. D. Meng, D. Z. Wang, J. H. Liu and S. Y. Zhang, *Mater. Res. Bull.*, 2004, **39**, 2163.
19. X. Sun, X. Chen and Y. Li, *Inorg. Chem.*, 2002, **41**, 4996.
20. A. R. Armstrong, G. Armstrong, J. Canales and P. G. Bruce, *Angew. Chem., Int. Ed.*, 2004, **43**, 2286.
21. S. Pavasupree, Y. Suzuki, S. Yoshikawa and R. Kawahata, *J. Solid State Chem.*, 2005, **178**, 3110.
22. G. Mogilevsky, Q. Chen, H. Kulkarni, A. Kleinhammes, W. M. Mullins and Y. Wu, *J. Phys. Chem. C*, 2008, **112**, 3239.
23. F. Alvarez-Ramirez and Y. Ruiz-Morales, *Chem. Mater.*, 2007, **19**, 2947.
24. J. Zhao, X. Wang, T. Sun and L. Li, *Nanotechnology*, 2005, **16**, 2450.
25. O. K. Varghese, D. Gong, M. Paulose, C. A. Grimes and E. C. Dickey, *J. Mater. Res.*, 2003, **18**, 156.
26. A. Ghicov, H. Tsuchiya, J. M. Macak and P. Schmuki, *Phys. Stat. Sol. A*, 2006, **203**, R28.
27. J. M. Macak, S. Aldabergerova, A. Ghicov and P. Schmuki, *Phys. Stat. Sol. A*, 2006, **203**, R67.
28. S. K. Pradhan, Y. Mao, S. S. Wong, P. Chupas and V. Petkov, *Chem. Mater.*, 2007, **19**, 6180.
29. T. Kubo and A. Nakahira, *J. Phys. Chem. C*, 2008, **112**, 1658.
30. J. Yu and M. Zhou, *Nanotechnology*, 2008, **19**, 045606.
31. H. Niu, Y. Cai, Y. Shi, F. Wei, S. Mou and G. Jiang, *J. Chromatogr., A*, 2007, **1172**, 113.
32. X. Sun and Y. Li, *Chem. Eur. J.*, 2003, **9**, 2229.
33. E. Morgado, M. A. S. de Abreu, G. T. Moure, B. A. Marinkovic, P. M. Jardim and A. S. Araujo, *Chem. Mater.*, 2007, **19**, 665.
34. D. V. Bavykin and F. C. Walsh, *J Phys. Chem. C*, 2007, **111**, 14644.
35. S. J. Gregg and K. S. W. Sing, *Adsorption, Surface Area and Porosity*, Academic Press, London, 2nd edn., 1982.
36. D. V. Bavykin, V. N. Parmon, A. A. Lapkin and F. C. Walsh, *J. Mater. Chem.*, 2004, **14**, 3370.
37. N. Wang, X. Li, Y. Wang, X. Quan and G. Chen, *Chem. Eng. J.*, 2009, **146**, 30.
38. M. Kruk, M. Jaroniec and A. Sayari, *Langmuir*, 1997, **13**, 6267.
39. T. Miyata, A. Endo, T. Ohmori, T. Akiya and M. Nakaiwa, *J. Colloid Interface Sci.*, 2003, **262**, 116.

40. X. Zhang, W. Wang, J. Chen and and Z. Shen, *Langmuir*, 2003, **19**, 6088.
41. D. V. Bavykin, E. V. Milsom, F. Marken, D. H. Kim, D. H. Marsha, D. J. Riley, F. C. Walsh, K. H. El-Abiary and A. A. Lapkin, *Electrochem. Commun.*, 2005, **7**, 1050.
42. L. Yezek, R. L. Rowell, L. Holysz and E. Chibowski, *J. Coll. Interface Sci.*, 2000, **225**, 227.
43. A. Fujishima, K. Hashimoto and T. Watanabe, *TiO2 photocatalysis: Fundamentals and Applications*, Bks, Inc, USA, 1999.
44. K. M. Glassford and J. R. Chelikowsky, *Phys. Rev. B*, 1992, **46**, 1284.
45. Y. Li, T. J. White and S. E. Lim, *J. Solid State Chem.*, 2004, **177**, 1372.
46. K. M. Reddy, C. V. G. Reddy and S. V. Manorama, *J. Solid State Chem.*, 2001, **158**, 180.
47. C. Kormann, D. W. Bahnemann and M. R. Hoffmann, *J. Phys. Chem.*, 1988, **92**, 5196.
48. N. Serpone, D. Lawless and R. Khairutdinov, *J. Phys. Chem.*, 1995, **99**, 16646.
49. T. Sasaki and M. Watanabe, *J. Phys. Chem. B*, 1997, **101**, 10159.
50. T. Sasaki, *Supramolec. Sci.*, 1998, **5**, 367.
51. N. Sakai, Y. Ebina, K. Takada and T. Sasaki, *J. Am. Chem. Soc.*, 2004, **126**, 5851.
52. H. Sato, K. Ono, T. Sasaki and A. Yamagishi, *J. Phys. Chem. B*, 2003, **107**, 9824.
53. D. V. Bavykin, S. N. Gordeev, A. V. Moskalenko, A. A. Lapkin and F. C. Walsh, *J. Phys. Chem. B*, 2005, **109**, 8565.
54. C. Kittel, *Introduction to Solid State Physics*, Wiley, New York, 8th edn, 2005.
55. J. R. Hook and H. E. Hall, *Solid State Physics*, Wiley, New York, 2003.
56. J. Hong, J. Cao, J. Sun, H. Li, H. Chen and M. Wang, *Chem. Phys. Lett.*, 2003, **380**, 366.
57. H. Xin, R. Ma, L. Wang, Y. Ebina, K. Takada and T. Sasaki, *Appl. Phys. Lett.*, 2004, **85**, 4187.
58. J. J. Yang, Z. S. Jin, X. D. Wang, W. Li, J. W. Zhang, S. L. Zhang, X. Y. Guo and Z. J. Zhang, *Dalton Trans.*, 2003, **20**, 3898.
59. J. Pascual, J. Camassel and H. Mathieu, *Phys. Rev. B: Condens. Matter*, 1978, **18**, 5606.
60. D. S. Boudreaux, F. Williams and A. J. Nozik, *J. Appl. Phys.*, 1980, **51**, 2158.
61. J. J. Kasinski, L. Gomez-Jahn, S. M. Gracewski, K. J. Faran and R. J. D. Miller, *J. Chem. Phys.*, 1989, **90**, 1253.
62. J. M. Cho, W. J. Yun, J. K. Lee, H. S. Lee, W. W. So, S. J. Moon, Y. Jia, H. Kulkarni and Y. Wu, *Appl. Phys. A*, 2007, **88**, 751.
63. M. Zhang, Z. Jin, J. Zhang, X. Guo, J. Yang, W. Li, X. Wang and Z. Zhang, *J. Mol. Catal. A: Chem.*, 2004, **217**, 203.
64. R. D. Iyengar, M. Codell, J. S. Karra and J. Turkevich, *J. Am. Chem. Soc.*, 1966, **88**, 5055.

65. M. A. Cortes-Jacome, G. Ferrat-Torres, L. F. Flores-Ortiz, C. Angeles-Chavez, E. Lopez-Salinas, J. Escobar, M. L. Mosqueira and J. A. Toledo-Antonio, *Catal. Today*, 2007, **126**, 248.
66. H. Berger, H. Tang and F. Levy, *J. Cryst. Growth*, 1993, **130**, 108.
67. G. S. Kim, S. G. Ansari, H. K. Seo, Y. S. Kim and H. S. Shina, *J. Appl. Phys.*, 2007, **101**, 024314.
68. H. K. Seo, G. S. Kim, S. G. Ansari, Y. S. Kim, H. S. Shin, K. H. Shim and E. K. Suh, *Solar Energy Mater. Solar Cells*, 2008, **92**, 1533.
69. D. V. Bavykin, M. Carravetta, A. N. Kulak and F. C. Walsh, in preparation.
70. E. Morgado Jr, M. A. S. Abreu, O. R. C. Pravia, B. A. Marinkovic, P. M. Jardim, F. C. Rizzo and A. S. Araujo, *Solid State Sci.*, 2006, **8**, 888.
71. W. Wang, J. Zhang, H. Huang, Z. Wu and Z. Zhang, *Colloids Surf. A*, 2008, **317**, 270.
72. O. P. Ferreira, A. G. S. Filho, J. M. Filho, O. L. Alves and J. Braz, *Chem. Soc.*, 2006, **17**(2), 393.
73. D. V. Bavykin, J. M. Friedrich, A. A. Lapkin and and F. C. Walsh, *Chem. Mat.*, 2006, **18**, 1124.
74. B. D. Yao, Y. F. Chan, X. Y. Zhang, W. F. Zhang, Z. Y. Yang and N. Wang, *Appl. Phys. Lett.*, 2003, **82**(2), 281.
75. L. Qian, Z. L. Du, S. Y. Yang and Z. S. Jin, *J. Mol. Struct.*, 2005, **749**, 103.
76. A. M. Rao, E. Richter, S. Bandow, B. Chase, P. C. Eklund, K. A. Williams, S. Fang, K. R. Subbaswamy, M. Menon, A. Thess, R. E. Smalley, G. Dresselhaus and M. S. Dresselhaus, *Science*, 1997, **275**, 187.
77. A. Thorne, A. Kruth, D. Tunstall, J. T. S. Irvine and W. Zhou, *J. Phys. Chem. B*, 2005, **109**, 5439.
78. T. Dittrich, J. Weidmann, V. Y. Timoshenko, A. A. Petrov, F. Koch, M. G. Lisachenko and E. Lebedev, *Mater. Sci. Eng., B*, 2000, **489**, 69.
79. J. R. Smith, F. C. Walsh and R. L. Clarke, *J. Appl. Electrochem.*, 1998, **28**, 1021.
80. H. Tokudome and M. Miyauchi, *Angew. Chem.*, 2005, **117**, 2010.
81. C. Damesa, B. Poudel, W. Z. Wang, J. Y. Huang, Z. F. Ren, Y. Sun, J. I. Oh, C. Opeil, M. J. Naughton and G. Chen, *Appl. Phys. Lett.*, 2005, **87**, 031901.
82. H. Chen, W. Yang, Y. He, Y. Ding, L. Zhang, C. Tan, A. A. Lapkin and D. V. Bavykin, *Powder Technol.*, 2008, **183**, 63.
83. G. K. Mor, O. K. Varghese, M. Paulose, K. Shankar and C. A. Grimes, *Sol. Energy Mater. Sol. Cells*, 2006, **90**, 2011.

CHAPTER 4

Chemical Properties, Transformation and Functionalization of Elongated Titanium Oxide Nanostructures

Bulk crystalline TiO_2 is a relatively durable compound, which tends to react only with strong acids or bases under heating. In contrast, nanostructured TiO_2 and protonated titanates tend to be much more reactive due to their increased surface area. This chapter reviews the known chemical (including ion-exchange and surface) properties of elongated nanostructured titanates and TiO_2. The thermodynamic aspects of the dispersed phase and stability of nanostructured materials are also addressed.

4.1 Thermodynamic Equilibrium between the Nanotube and its Environment

The interaction of nanostructured materials with their surrounding environment occurs at their interface. In comparison with bulk materials, nanostructures have a reduced size and an increased surface area, which can lead to such interactions having a significant impact. This results in an alteration in the chemical and physico-chemical properties of nanostructures, and these properties may become size dependent.

The dissolution of nanostructured TiO_2 and titanate in aqueous solutions can be described using three possible chemical reactions:

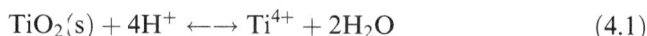

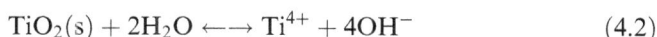

$$TiO_2(s) + 4H^+ \longleftrightarrow Ti^{4+} + 2H_2O \tag{4.1}$$

$$TiO_2(s) + 2H_2O \longleftrightarrow Ti^{4+} + 4OH^- \tag{4.2}$$

RSC Nanoscience & Nanotechnology No. 12
Titanate and Titania Nanotubes: Synthesis, Properties and Applications
By Dmitry V. Bavykin and Frank C. Walsh
© Dmitry V. Bavykin and Frank C. Walsh 2010
Published by the Royal Society of Chemistry, www.rsc.org

$$\mathrm{TiO_2(s) + 2OH^- \longleftrightarrow TiO_3^{2-} + H_2O} \tag{4.3}$$

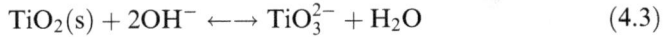

The exact form of the $\mathrm{Ti^{4+}}$ and $\mathrm{TiO_3}^{2-}$ ions may be more sophisticated, including the formation of soluble hydroxides or polytitanic anions, however, for demonstration purposes their simplified form is used here. Depending on the pH of the solution, one of these reactions will dominate dissolution.

At thermodynamic equilibrium, the chemical potentials of each species, μ_i obey the following equation:

$$\sum_i \nu_i \mu_i = 0 \tag{4.4}$$

where ν_i values are the stochiometric coefficients in reactions (4.1), (4.2) and (4.3), which are positive for reactants and negative for products. The chemical potential of dissolved components can be expressed as:

$$\mu_i = \mu_i^0 + RT \ln(X_i) \tag{4.5}$$

where R is the Boltzmann constant, T is the temperature and X_i is the molar fraction of component i in solution, and μ_i^0 is the chemical potential under standard conditions.

The chemical potential of solid $\mathrm{TiO_2}$ may be calculated by recalling the definition of chemical potential as a partial derivative of thermodynamic potential G_{TiO_2} over the amount of $\mathrm{TiO_2}$, *i.e.* n_{TiO_2}. Since a change in the amount (number of moles) in solid $\mathrm{TiO_2}$ results in a variation in its surface area, A, the chemical potential will become equal to:

$$\mu_{TiO_2} = \left(\frac{\partial G_{TiO_2}}{\partial n_{TiO_2}}\right)_{T,P,n_i} = \left(\frac{\partial G^b_{TiO_2}}{\partial n_{TiO_2}}\right)_{T,P,n_i} + \sigma V_m \left(\frac{dA}{dV}\right)$$
$$= \mu^b_{TiO_2} + \sigma V_m \left(\frac{dA}{dV}\right) \tag{4.6}$$

where $\mu^b_{TiO_2}$ is the chemical potential of $\mathrm{TiO_2}$ without an interfacial boundary (infinite crystal), σ is the excess surface energy in the $\mathrm{TiO_2}$–solution interface, V_m is the partial molar volume of $\mathrm{TiO_2}$, A and V are the surface area and volume of the $\mathrm{TiO_2}$ nanostructure, respectively. Let us consider the change in the surface area, A, and volume, V, during the dissolution–crystallization of nanotubes, providing that this process does not occur on the ends of nanotubes, but rather on the nanotube wall surfaces. Equation (4.6) is only valid if the surface energy per unit area, σ, and the molar volume, V_m, are constant even when the nanostructures are small.

In nanotubes there are two types of surface, namely the internal concave surface, A_{int}, and the external convex surface, A_{ext}. Deposition of materials on

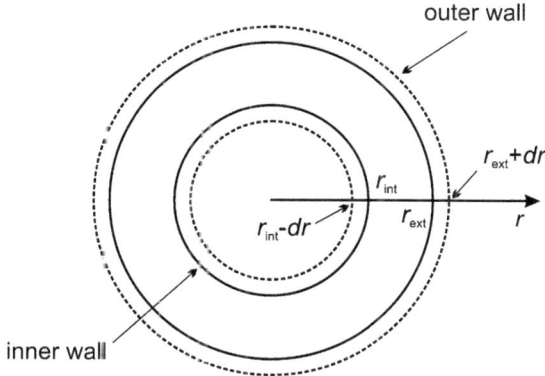

Figure 4.1 Diagram showing the change in nanotube radius during crystallization onto the walls.

both sides of the nanotubes, results in an incremental change in the internal radius, r_{int}, and the external radius, r_{ext}, of the nanotubes (see Figure 4.1).

Providing this change is differentially small, it is possible to express the last part of Equation (4.6) as:

$$\frac{dA}{dV} = \frac{dA_{ext}}{dV_{ext}} + \frac{dA_{int}}{dV_{int}} = \frac{2\pi L_{tube} dr}{2\pi L_{tube} r_{ext} dr} - \frac{2\pi L_{tube} dr}{2\pi L_{tube} r_{int} dr} = \frac{1}{r_{ext}} - \frac{1}{r_{int}} \quad (4.7)$$

where L_{tube} is the nanotube length. Inserting Equation (4.7) into (4.6) provides a formula for the chemical potential of solid nanotubes:

$$u_{TiO_2} = \mu_{TiO_2}^b + \sigma V_m \left(\frac{1}{r_{ext}} - \frac{1}{r_{int}} \right) \quad (4.8)$$

Substitution of Equation (4.8) and (4.5) into Equation (4.4), followed by a rearrangement leads to:

$$\sigma V_m \left(\frac{1}{r_{ext}} - \frac{1}{r_{int}} \right) + \sum_i \nu_i \mu_i^0 = RT \sum_i (-\nu_i) \ln X_i \quad (4.9)$$

Taking the exponential of Equation (4.9) and performing a rearrangement, results in a formula for the solubility product of TiO_2 nanotubes:

$$K_{sp} = \prod_i X_i^{-\nu_i} = K_{sp}^b e^{\frac{\sigma V_m}{RT} \left(\frac{1}{r_{ext}} - \frac{1}{r_{int}} \right)} \quad (4.10)$$

where K_{sp}^b is the solubility product of bulk TiO_2 with a flat surface (*i.e.*, zero-curvature). The solubility of titanate nanotubes is size dependent.

The equilibrium concentration of titanium(IV) near an external (convex) surface is higher than that near an internal (concave) surface in accordance with Equation (4.10), resulting in a thermodynamic metastability of nanotubular morphology. Indeed, in the absence of activation barriers to dissolution–crystallisation, there would be a continuous re-crystallisation of solid materials from external convex to internal concave surfaces due to a difference in solubility, resulting in the collapse of internal hollow cavities leading to the formation of nanorods. The mechanism of this process is similar to that of the process of Ostwald ripening.[1] It is possible that a layered structure of titanate nanotubes can provide a higher activation barrier towards dissolution from and crystallisation onto the walls of the nanotubes, resulting in a stabilisation of the metastable nanotubular state.

Topologically, nanotubes are characterised by a larger r_{ext} and a smaller r_{int}, resulting in the overall nanotube solubility product, K_{sp}, as calculated from Equation (4.10), always being smaller than that of bulk K_{sp}^b.

In general, any first-order phase transformation process with a two-phase mixture composed of a dispersed second phase in a matrix, is characterised by an excess of free energy associated with the interface, as seen in Equation (4.6). As a result, many properties of the material become dependent on the size of the nanoparticles. Since underlying phenomena resulting in such size-dependence are common, the resulting equations have similar forms. For example, the vapour pressure, p, above the liquid, depends on the two principal radii of interface curvature, r_1 and r_2, according to the Kelvin equation:[2]

$$\ln\left(\frac{p}{p_0}\right) = \frac{2\sigma_{g-l}V_m}{RT}\left(\frac{1}{r_1}+\frac{1}{r_2}\right) \tag{4.11}$$

where p_0 is the saturation vapour pressure at temperature T, σ_{g-l} is the interfacial tension at the gas–liquid interface, and V_m is the molar volume of the liquid. There is a similarity between Equations (4.10) and (4.11).

A decrease in the size of solid particles can also result in a melting-point depression phenomenon, when the melting temperature, T_m, of nanosized crystals becomes size dependent. Such a dependence can be described using the Gibbs–Thomson equation:[3]

$$T_m = T_m^b\left(1 - \frac{4\sigma}{H_m\rho_s d}\right) \tag{4.12}$$

where T_m^b is the melting temperature of an infinite bulk crystal with a flat surface, H_m is the bulk latent heat of fusion (melting), ρ_s is the density of solid particles, σ is the excess free surface energy per unit area, and d is the diameter of the spheroidal particles. Equation (4.12) demonstrates a similar dependence on $1/d$ as Equations (4.11) and (4.10), suggesting a common underlying thermodynamic effect of the interface on the processes of melting, evaporation and dissolution of the dispersed phases.

4.2 Ion-Exchange Properties of Nanostructured Titanates

Titanate nanotubes belong to the class of weak acids, whose general formula can be represented as $H_2Ti_nO_{2n+1} \cdot xH_2O$. It has been shown[4] that exchangeable cations occupy positions between the titanate layers (inside the walls of multi-wall nanotubes), as well as on the surface of the nanotubes. In solid nanofibres, the cations of alkali metals also occupy the interlayer spacing between (100) planes inside the fibre. The reaction of proton exchange with alkali metal cations in an aqueous suspension of nanotubes or nanofibres can, therefore, be represented as follows:

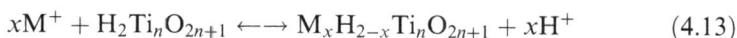

$$xM^+ + H_2Ti_nO_{2n+1} \longleftrightarrow M_xH_{2-x}Ti_nO_{2n+1} + xH^+ \qquad (4.13)$$

where M^+ is the alkaline metal cation. The ion-exchange reaction (4.13) is reversible, and the value of x is expected to depend on pH, M^+ concentration and temperature.

The degree of ion-exchange can be very high for singly-charged alkaline ions, with the value of x approaching 2 (ref. 5), whereas double- or triply-charged cations have reduced ability to substitute protons,[6] probably due to the internal stress which results from charge compensation.

4.2.1 Kinetic Characteristics of Ion-Exchange

Nanostructured titanates, produced by the alkaline hydrothermal treatment of TiO_2, are characterized by an open, mesoporous morphology, which facilitates the transport of ions from a liquid towards the surface of nanotubes. Figure 4.2 shows the dynamics of pH change in an aqueous suspension of nanostructured titanate and titania after the addition of aqueous LiOH. For a blank solution of pure water, the addition of LiOH results in a very rapid increase in pH up to a value of 11.5, with an almost constant value after 60 min. A small decrease of pH is attributable to the reaction with atmospheric CO_2, which leads to the formation of lithium carbonate. The addition of an identical amount of LiOH to a suspension of TiO_2 nanoparticles (Degussa P-25; characterized by spheroidal particles of *ca.* 20 nm diameter), results in a rapid increase in pH up to 11, followed by an insignificant drop. In contrast, the addition of an identical amount of LiOH to a suspension of titanate nanotubes and nanofibres, results in an initial, rapid rise in pH to approximately 11, followed by a slow decrease in pH to the values 7.3 and 8.1, respectively. The characteristic time for the pH decrease is in the range of several tens of minutes.

The specific surface area of Degussa P-25 (*ca.* 50 m^2 g^{-1}) is lower than that of titanate nanotubes[7] (*ca.* 200 m^2 g^{-1}), but higher than that of titanate nanofibres (*ca.* 20 m^2 g^{-1}). Thus, such an insignificant drop in pH after the addition of alkali to P-25 cannot simply be explained in terms of a small surface area. Rather, such a high degree of pH variation for nanostructured titanates can be explained by the leaching of protons from the crystal structure of the

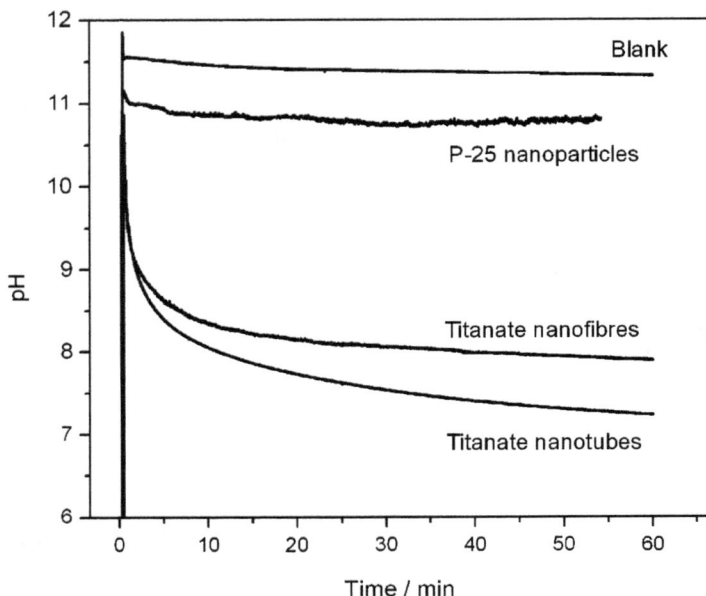

Figure 4.2 Graph showing pH as a function of time, following the addition of 50 μL of LiOH ($1 \, mol \, dm^{-3}$) to a suspension of 0.1 g of TiO_2 nanoparticles (P-25), titanate nanotubes or titanate nanofibres in 10 mL water at 25 °C. The initial pH of water is *ca.* 4.5. (Data are adapted from ref. 8).

protonated titanates, which arises from the substitution of protons by lithium cations, as shown in reaction (4.13). The dynamics of pH change reflect the kinetic regularities of ion-exchange in nanotubular and nanofibrous titanates.

The pH of a solution is related to the concentration of protons using a logarithm, $pH = -\log([H^+])$. The typical kinetic curve of proton concentration growth in a suspension of protonated titanate nanotubes after addition of LiOH, had the "S"-shape shown in Figure 4.3. The curve indicates a short induction period, followed by monotonic growth and saturation. The release of protons from the titanate nanotubes to the bulk solution can be illustrated using the following sequence of processes:

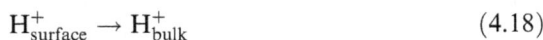

$$Li^+_{bulk} \rightarrow Li^+_{surface} \tag{4.14}$$

$$Li^+_{surface} \rightarrow Li^+_{intercalate} \tag{4.15}$$

$$Li^+_{intercalate} + H_2Ti_nO_{2n+1} \rightarrow LiHTi_nO_{2n+1} + H^+_{intercalate} \tag{4.16}$$

$$H^+_{intercalate} \rightarrow H^+_{surface} \tag{4.17}$$

$$H^+_{surface} \rightarrow H^+_{bulk} \tag{4.18}$$

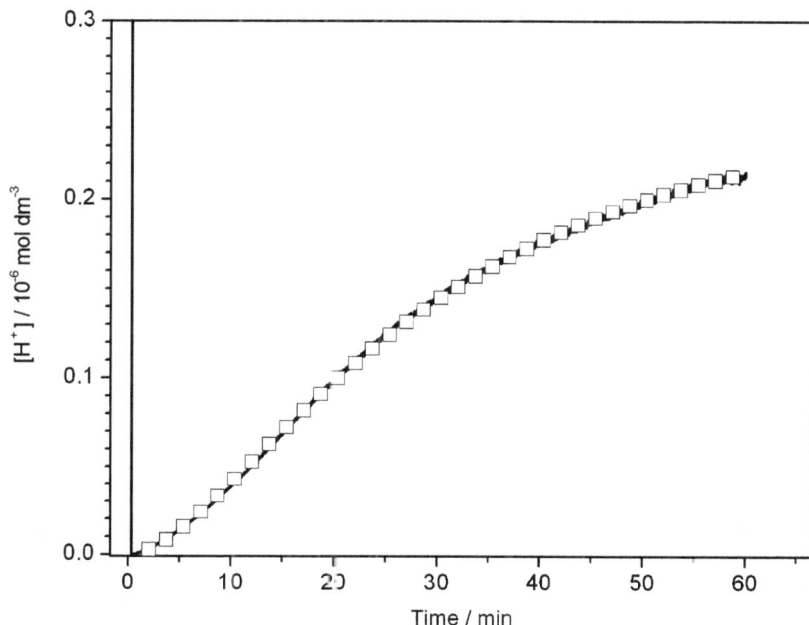

Figure 4.3 Graph showing proton concentration growth with time ($\square$) after addition of 40 mL of LiOH (1 mol dm^{-3}) to a suspension of 0.1 g of protonated titanate nanotubes in 10 mL water at 25 °C. The line was fitted using Equation (4.20). (Data are adapted from ref. 8).

The resultant process corresponds to reaction (4.13), with Li$^+$ as the inter-calating ion and $x = 1$. Step (4.14) represents the adsorption of lithium ions from bulk solution onto the surface of the nanotubes, whilst step (4.15) is the transport of lithium ions inside the crystal to the ion-exchange centres. Reaction (4.16) represents the ion–exchange which occurs inside the crystal. Steps (4.17) and (4.18) represent the reverse transport of protons from the crystal to the surface, then into the bulk solution. It has been shown that the length of nanotubes affects the overall rate of ion-exchange.[8] This may indicate that the transport of lithium ions and protons between the layers inside the nanotube wall are the rate-limiting steps. The above scheme can therefore be simplified by the exclusion of fast processes in two consecutive stages, namely the diffusion of lithium ions (A to B) and the diffusion of protons (B to C) inside the nanotube walls:

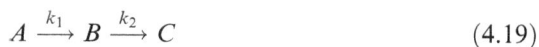

$$A \xrightarrow{k_1} B \xrightarrow{k_2} C \tag{4.19}$$

where k_1 and k_2 are reaction constants. Scheme (4.19) is deliberately over-simplified in order to demonstrate the appearance of an induction time in the kinetic curve of proton concentration growth. More detailed treatments should consider the transformation from A to B, not as a chemical reaction, but rather as a diffusion process inside the nanotube. The analytical solution for the accumulation of C from scheme (4.19), using the law of mass action, has the form:[9]

$$C(t) = \frac{A_0}{k_2 - k_1}\left(k_2\left(1 - e^{-k_1 t}\right) - k_1\left(1 - e^{-k_2 t}\right)\right) \qquad (4.20)$$

where A_0 is the infinity concentration of protons, and k_1 and k_2 are parameters characterising the diffusion of lithium ions and protons inside the nanotubes, respectively.

Equation (4.20) is fitted to the experimental kinetic curve of proton concentration growth in Figure 4.3. The proposed scheme (4.19) is in agreement with experimental kinetics and explains the reasons for the induction period in the kinetic curve. The growth in proton concentration in suspension following the addition of strong LiOH, is due to a leaching of protons from the solid nanotubes as a result of ion exchange with the lithium cations diffused inside the nanotubes. At zero time, the concentration of lithium ions inside the nanotubes and the rate of the proton leaching are both equal to zero. As soon as lithium ions diffuse into the nanotubes, the reverse leaching of protons occurs after an induction time.

Although the mechanistic scheme (4.19) does not correspond exactly to the process of ion diffusion in solid materials described, a general solution for diffusion in a cylinder[10] can be used to show the relation of the reaction constant k_1 to the diffusion coefficient:

$$k_1 \sim \frac{D_{Li}}{L_{tube}^2} \qquad (4.21)$$

where D_{Li} is the diffusion coefficient of lithium ions inside the titanate nanotubes, and L_{tube} is the nanotube length.

An analysis of k_1 reaction constants for samples of nanotubes having differing average lengths, shows that Equation (4.21) is valid.[8] The values of k_1 were obtained by fitting experimental kinetic curves describing proton concentration growth for nanotubes of differing lengths (see Figure 4.3) to Equation (4.20). Average values for nanotube length were used, with determination using SEM microscopy. The determined values for k_1 and L_{tube} are consistent with Equation (4.21), indicating that the transport of lithium ions in titanate nanotubes probably occurs along the length of the nanotube. The value of the diffusion coefficient was estimated in the order of magnitude of 10^{-11} $cm^2 s^{-1}$, which is much smaller than the diffusion coefficient of caesium ions in zeolite, but of a similar order of magnitude to that of *cancrinite*.[11] Similar diffusion coefficient values were also observed for zirconium thiophophates.[12] For a more accurate determination of the diffusion coefficient, an improved model for the forward and reverse diffusion of ions in cylindrical particles is required.

Nanofibrous titanate nanostructures demonstrate similar ion-exchange properties to those of titanate nanotubes. Although the amount of ion-exchangeable centres in nanofibres is less than that in nanotubes, the characteristic time of ion-exchange between protons and lithium ion from aqueous solution is of the same order of magnitude (see Figure 4.2). Studies of the kinetic regularities of alkaline metal ion intercalation into nanofibrous titanates of differing length and width, have shown that changes in the dimensions of nanofibres facilitate ion-exchange without changing the extent of the process, confirming that the ion transport inside nanofibres is a rate-limiting stage.[8]

In addition, crystallographic studies of intercalated alkaline ions inside titanate nanotubes and nanofibres have clarified the directions of ion flow in nanostructures. Figure 4.4 shows the transformation of titanate nanosheets to nanotubes or nanofibres, with a consistently oriented crystallographic axis. During the growth of nanosheets, under alkaline hydrothermal conditions, the bending of nanosheets around axis b occurs followed by a closing of the loop, and growth along direction b. If the formation of a nanotube does not occur, the nanofibres produced have a length corresponding the direction c, and a thickness corresponding to both directions a and b (see Figure 4.4). As described above, the transport of alkaline metal cations in titanate nanotubes occurs preferentially along the nanotube length, which corresponds to axis b. In

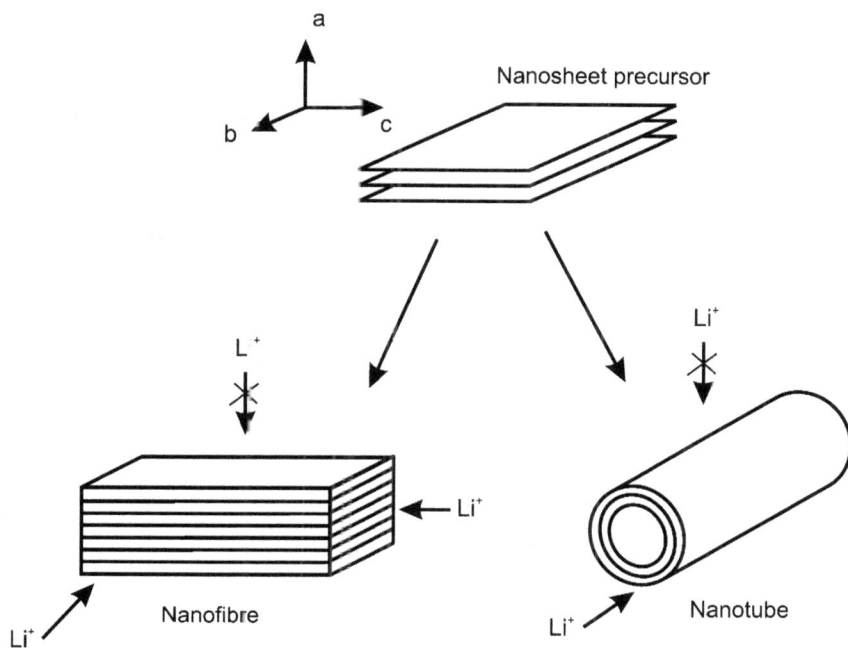

Figure 4.4   Titanate nanostructures (nanosheets, nanofibres and nanotubes with crystallographic axis) and the directions of cation transport during ion-exchange.

the case of nanofibres, the transport of cations occurs preferentially along axes b and c. The length of the nanotubes (in the hundreds of nanometres range) is of the same order of magnitude as the thickness of nanofibres (along direction b). Taking into account that the characteristic time of ion-exchange for both nanotubes and nanofibres is similar, it may be concluded that the transport of ions in nanofibres in direction b is dominant.

The diffusion of alkaline metal ions inside nanotubes occurs between the titanate layers, which are characterized by a zigzag structure and an average spacing of *ca.* 0.72 nm. The sizes of the alkaline metal ions in aqueous solution are smaller and have the following ionic diameters: Li^+ (0.09 nm), Na^+ (0.116 nm), K^+ (0.152 nm), Cs^+ (0.181 nm) for a coordination number of 6 (ref. 13). Despite the differences in the size of alkali metal cations, the apparent rate of intercalation is very similar for all of the ions studied. This unusual result of the absence of selectivity of titanate nanotubes in terms of alkali metal transport is consistent with the observation of crystal structure alterations following the substitution of protons by alkali metal cations in nanotubular titanates, showing an increase of interplanar distance d_{200} from 0.88 nm to 1.18 nm.[8] Surprisingly, the interplanar distance $d_{200} = 1.18$ nm is almost identical for many of the nanotubular titanates of alkaline metals, including: Li^+, Na^+, K^+ and Cs^+.

4.2.2 Decoration of Nanotubes using the Ion-Exchange Method

The ion-exchange properties of nanostructured titanates can be utilized in the deposition of the dispersed nanoparticles of other materials onto the surface of titanates. Such an approach is schematically illustrated in Figure 4.5. The precursor of the required nanoparticles in cationic form is ion-exchanged with the protons of the nanostructured titanates, which are localised on both the concave and convex nanotube surfaces, as well as in the interlayer space in the nanotube wall. Ideally, such ion-exchange processes should allow a uniform atomic distribution of the precursor on the nanotube surfaces by adsorption, as well as between the layers of the walls by intercalation. The overall process follows chemical reaction (4.13), where M^+ is the cationic precursor. Such ion-exchanged materials are thoroughly washed with solvent to remove the impurities of precursors retained in the pores between the nano-tubes, preventing the formation of nanoparticles in the bulk material during next step.

The second step of the procedure is a growth of nanoparticles of the required materials on the nanotube surface, using a chemical reaction between the intercalated precursor and added chemical reagents. During this stage, the precursor diffuses to the surface of nanotubes, where it reacts with chemicals added to the solution to form nanoparticles which adhere to the nanotube surface (see Figure 4.5). Due to the suppression of nanoparticle crystallisation in the bulk material and an agglomeration of particles on the surface, the composites obtained are characterised by a high loading of nanoparticles, which are evenly distributed on the surface of the titanate nanotubes.

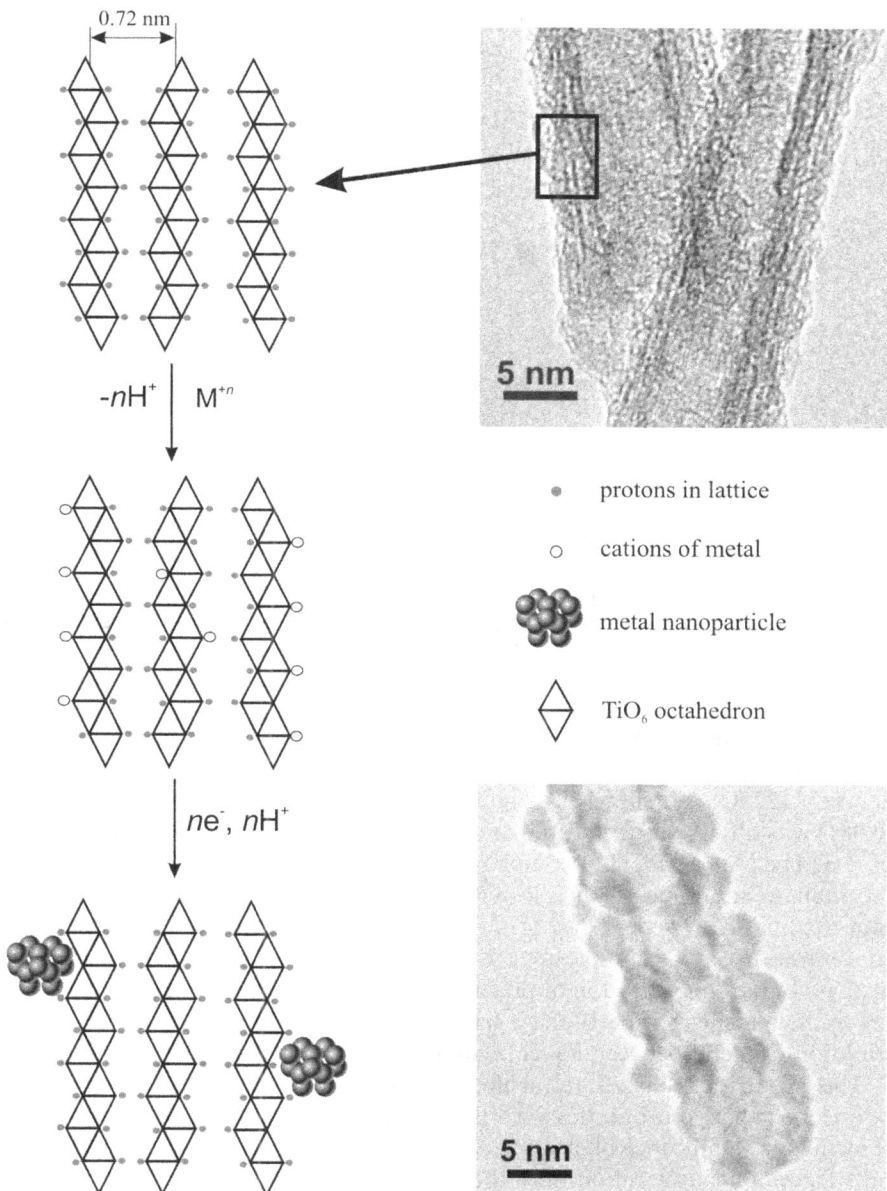

Figure 4.5 Ion-exchange assisted deposition of Pd nanoparticles onto the surface of titanate nanotubes.

The ionic form of the precursor is important for the successful distribution of nanoparticles and for a high material loading. The effect of speciation of the metal complex on its ability to participate in ion-exchange with titanate nanotubes, was studied using the the polyamine complexes of gold.[14] The adsorption

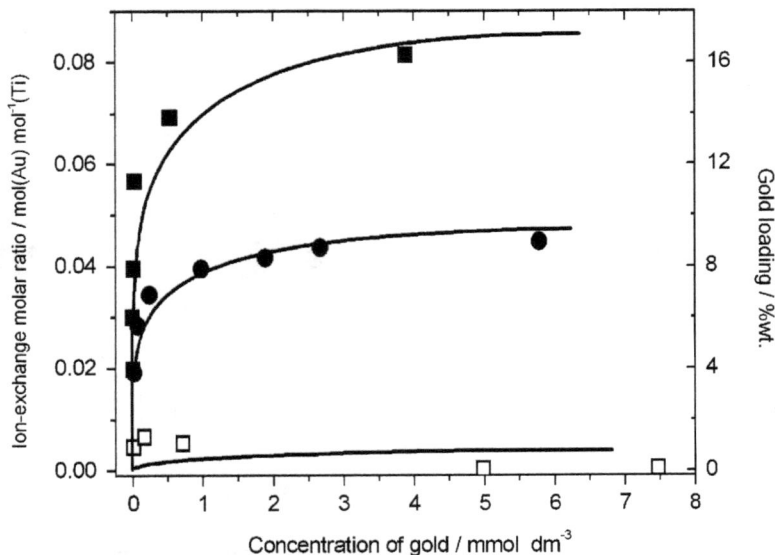

Figure 4.6   Isotherms for the adsorption of three gold complexes on titanate nanotubes from an aqueous suspension at 22 °C: (■) [Au(en)$_2$]Cl$_3$, (●)
[Au(dien)Cl]Cl$_2$ and (■) H[AuCl$_4$].

isotherms of three different complexes, namely [Au(en)$_2$]Cl$_3$, [Au(dien)Cl]Cl$_2$
and H[AuCl$_4$], from aqueous solution onto titanate nanotubes at room temperature (22 °C) are shown in Figure 4.6. The highest molar exchange ratio
(0.081) was observed for the gold complex, [Au(en)$_2$]Cl$_3$, a cation with a charge
of + 3. The [Au(dien)Cl]Cl$_2$ complex, which has a charge of + 2, adsorbs onto
the titanate nanotubes with a lower value for the maximum molar exchange
ratio (*ca.* 0.045). These values of molar exchange ratio are achieved when the
concentration of gold complexes is in the range of several mmol dm^{-3}. A further
increase in the concentration of precursor in solution results in saturation of the
ion-exchange sites in the titanate nanotubes, establishing an equilibrium distribution of metal between aqueous solution and the solid nanotubes.

The negatively-charged tetrachloride complex of gold demonstrates very
poor ion-exchange properties with the negatively-charged titanate nanotubes,
showing the significance of electrostatic interactions for the adsorption of metal
salts from aqueous solution onto the titanate nanotubes.

Suitable precursors include any species characterised by a strong affinity to
titanate nanotubes, such as: cations of precious metals,[15] transition metals,[16,17]
or nitrogen-containing organics.[18] The process of nanoparticle growth can be
stimulated by various chemical processes, including: the reduction of metal
cations to metal nanoparticles[14,15] (see Figure 4.7d), the sol–gel hydrolysis of
metal cations to oxides[19] (see Figure 4.7a), the recipitation of insoluble
solids[5,16] (see Figure 4.7c), or the oxidative polymerization of monomers[18] (see
Figure 4.7b).

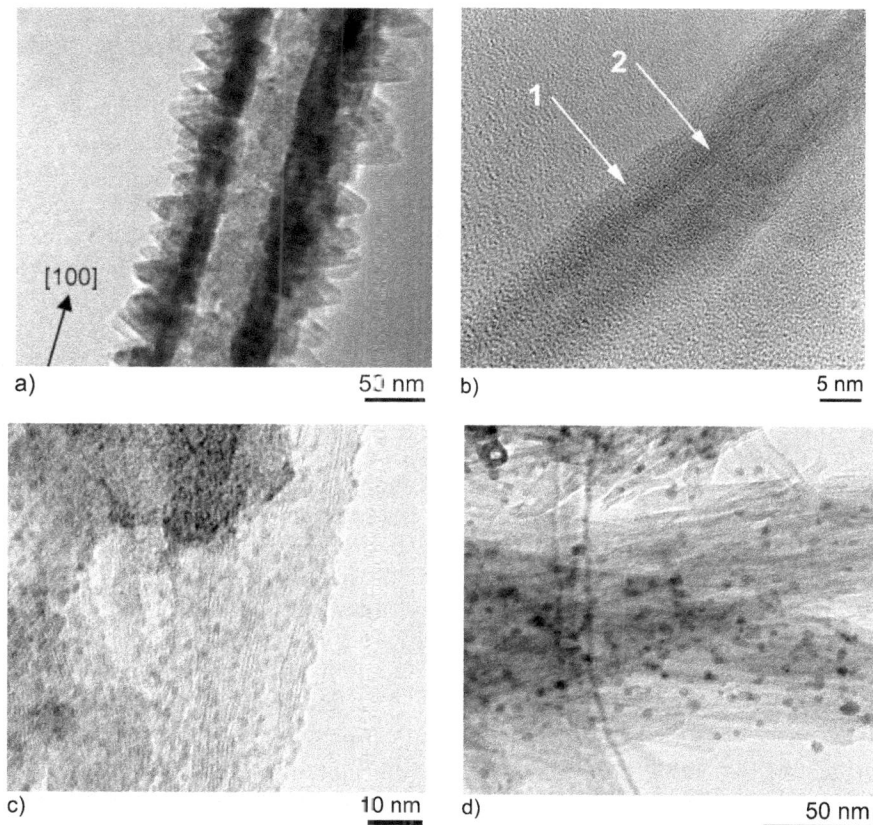

Figure 4.7 TEM images of nanostructured titanates decorated with coatings using the ion-exchange method: a) TiO$_2$/TiNF, b) PANI/TiNT, c) RuOOH/TiNT and d) Au/TiNT composites. In a) the arrow shows the crystallographic direction of the nanofibre, and in b) arrows 1 and 2 show the crystallographic directions of the polymer and nanotubes, respectively. (Images a) and b) are reproduced with kind permission from ref. 19 and 18, respectively).

The resultant composites are characterised by a good adhesion between the titanate substrate and coatings; a uniform distribution of the coating material on the surface of the nanostructured titanates; and a high dispersity of nanoparticles even at very high loading of supported materials (see Figure 4.7). This method of ion-exchange is perspective for preparation of dispersed catalysts on the surface of mesoporous nanotubular supports and particular examples are considered in Chapter 5.

4.2.3 Decoration of Substrates with Nanotubes

One of the key advantages of TiO$_2$ nanotube arrays produced by anodization of titanium is that they are already deposited as a film on the surface of the substrate, and can be consequently used immediately in various engineering applications. In contrast, the powdered titanate nanotubes require further processing for their immobilization on various supports.

During the coating of surfaces with nanostructured titanates, the following properties of the deposits are controlled: the adhesion to the substrate; the durability and hardness of the coating; the uniformity of material distribution; and the thickness and density of the films, as well as their composition. A particular challenge is the ability to control the orientation of nanotubes, to be either perpendicular or parallel relative to the substrate surface. Brief descriptions of various methods used in nanotube immobilisation follow.

The *doctor blade* technique is commonly used for the preparation of the nanotube composite electrodes used in electrochemical studies.[20,21] In general, the nanotube powder is dispersed in a minimal volume of solvent (typically 1–3 g in 10 cm^3 of solvent) using a shear blade mixer of an ultrasonic dispergator, to form a dense slurry. The slurry is mixed with dissolved polymer binder and any other optional additives (*e.g.* carbon nanoparticles for improving electro-conductivity). The prepared mixed slurry is deposited on the flat surface of the substrate using metal bars to ensure an even distribution of material, followed by evaporation of the solvent and a curing of the film at elevated temperature. The film thickness varies from a fraction of a millimetre to several millimetres.

A variation on the above method is the *spin coating* technique, in which the slurry is deposited into the centre of the spinning substrate, so that it flows towards the edge under centrifugal force. The film thickness, uniformity and quality are controlled by the rate of slurry introduction, the viscosity of the slurry and the spinning rate. Typically, the films produced are thinner than those prepared by the doctor blade technique.[22]

Titanate nanotubes can be immobilised onto the surface of the titanium substrate during their synthesis under alkaline hydrothermal conditions, using the method of *in situ growth*. In this technique, the metal titanium is oxidised to intermediate TiO_2, which undergoes further transformation to titanate nanotubes which are immobilised onto the surface of the dissolving titanium. The nanotubes produced are usually randomly oriented,[23] however, some reports[24] suggest a small degree of orientation in such films. The thickness of these films can reach the value of several hundreds of microns. Preliminary vacuum spattering of titanium onto the substrate, followed by the *in situ* alkaline hydrothermal growth of titanate nanotubes, also allows films to be deposited onto non titanium substrates. In this case, the choice of substrates is restricted by the durability of the support in the alkaline environment of the hydro-thermal conditions.

In an aqueous suspension, titanate nanotubes develop a relatively large negative zeta potential due to acid–base dissociation. Under a potential gradient, titanate nanotubes experience an electrophoretic motion towards a positively-charged electrode. Due to their close proximity to the electrode, the nanotube concentration increases resulting in precipitation onto the electrode surface.[25] The method of *electrophoretic deposition* (EPD) of ceramic nanoparticles onto the surface of an electrode, is based on this principle of particle

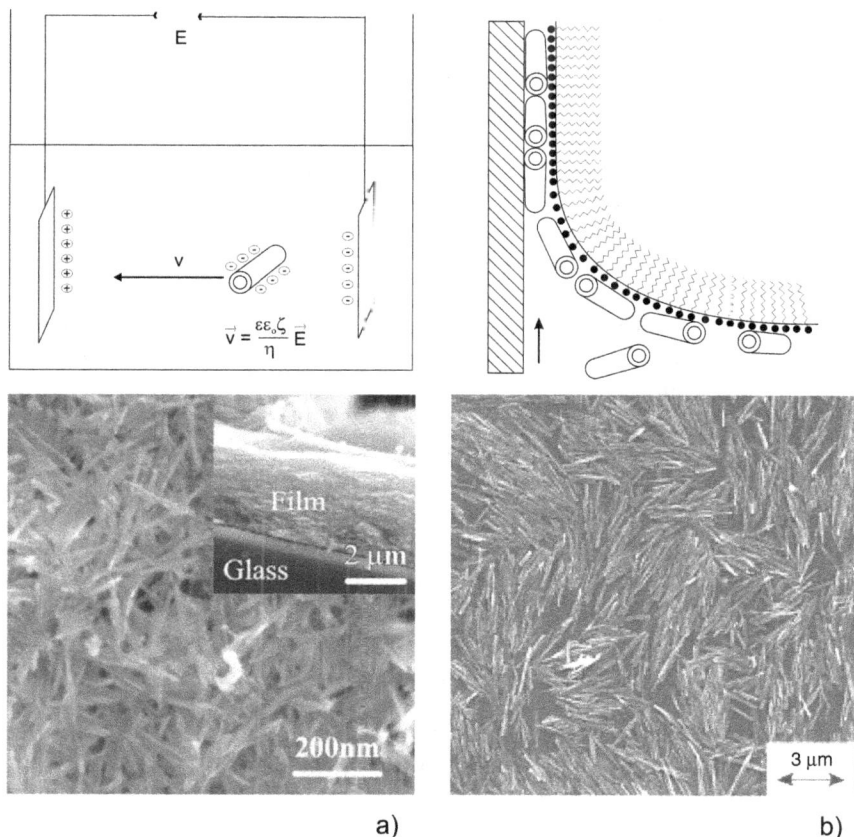

Figure 4.8 Schematic representations and SEM images of: a) electrophoretic, and b) Langmuir–Blodgett deposition of titanate nanotubes on a planar substrate. (SEM images a) and b) are reproduced from ref. 26 and 29, respectively).

migration, followed by their coagulation onto the electrode surface (see Figure 4.8a; ref. 26,27). The SEM image in Figure 4.8a, shows the typical morphology of a titanate nanotube film obtained by the EPD method. Note that the nanotubes are randomly oriented and packed in a dense film layer. The thickness of the layer depends on the time of EPD and can be greater than several microns. The films are often consolidated by subsequent heat treatment.

The typical electrolyte for EPD is the stable suspension of titanate nanotubes in alcohol or an alcohol–water mixture. During deposition, the distance between electrodes is minimized in order to increase the electric field, to typically a few centimetres. The EPD is carried out at a typical potential range of 10 to 40 V for a period of time of up to 60 min.[20,27] The addition of various stabilizing agents (*e.g.* polyvinyl butyral; M_w: 19 000,[28]), can result in a significant change in the zeta potential, even switching it to positive values and allowing tthe deposition of nanotubes on the surface of the cathode.

The *Langmuir–Blodgett technique* for film manufacture provides a better control of layer structure and the potential for nanotubes to self assemble with a preferential orientation. The method also enables the deposition of a monolayer of inorganic nanoparticles. The principle of this method is shown schematically in Figure 4.8b. Certain amphiphilic surfactant molecules are preferentially localised at the solvent–air interface, such that the hydrophilic part of the molecule is submerged in the liquid phase, whereas the hydrophobic part is removed from the liquid phase. The decrease in interface area usually results in the formation of a condensed layer with self-organised surfactant. If the hydrophilic part of the surfactant has a cationic form (*e.g.* cetyltrimethyl ammonium), it can attract negatively-charged titanate nanotubes to the surface. Slow withdrawal of the substrate allows nanotubular titanates to be trapped between substrate and surfactant layer, this can result in the formation of a monolayer of nanotubes.

Recent experiments have shown that a monolayer of titanate nanotubes can be formed on the surface of solid substrate by the Langmuir–Blodgett technique, using cetyltrimethyl ammonium chloride as surfactant.[29] An SEM image of a monolayer of nanotubes is shown in Figure 4.8b. The increase in surface pressure during deposition (from 1 to 18 mN m^{-1}), results in a dense monolayer film of nanotubes which consists of isolated grains with parallel tubes. The size of these grains is as large as several microns (see Figure 4.8b). The task of forming larger grains would require a narrower distribution of nanotube length and diameter.

For several applications using electron transport in titanate nanotubes, there would be the additional requirements of a perpendicular orientation of nanotubes relative to the substrate surface. This problem is very challenging, however, and requires a better understanding of the nanotube surface chemistry in order to successfully control the nanotube self-assembling process in the liquid phase.

4.3   Surface Chemistry and Functionalization of Nanostructured Titanates

The surface chemistry of nanostructured titanates is more versatile compared to the relatively inert chemistry of carbon nanotubes. The latter requires special treatment under aggressive conditions in order to activate their surface, using carboxylic groups to provide a flexible route for further functionalization. However, even in this case the surface density of carboxylic groups is relatively low. By contrast, nanotubular titanates are abundant with surface –OH groups, which are characterised by weak Bronsted acidity. The coverage density of surface –OH groups is estimated from the lattice parameters of trititanates, and taking into account that the nanotube surface corresponds to the (100) plane and that there are two –OH groups in the area of 0.375 x 0.919 nm^2 (see Table 3.1 in Chapter 3). These considerations suggest that a nanotube surface area of 1 nm^2 contains approximately 5.8 –OH groups.

The abundance of –OH groups on the surface of titanate nanotubes largely determines their chemical behaviour and the routes for their functionalization. One dramatic difference between titanate nanotubes and TiO_2 nanoparticles (such as P-25), is their reactivity with a H_2O_2 solution (30 wt%) at room temperature which results in the formation of titanium(IV) peroxo-complexes on the surface of the nanotubes, but not on that of the nanoparticles (see Figure 4.9). These peroxo-complexes are easily detected by a characteristic absorption band at 420 nm. Such surfaces, containing titanate nanotubes functionalized with surface peroxo-complexes, can be utilized in catalytic surface redox reactions.

The surface of titanate nanotubes is characterised by its hydrophilic properties due to a strong surface dipole moment and a high concentration of –OH groups. In order to modify the surface properties of the nanotubes to become hydrophobic, it is necessary to functionalise the surface, for example, by covering it with molecules containing long-alkane chains. The most popular hydrophobization method, adopted from silicon dioxide chemistry, is the controlled hydrolysis of trialkoxysilanes (*e.g.* allyltriethoxysilane[30]) on the nanostructure surface in anhydrous solvents (see Figure 4.9). Strong covalent Ti–O–Si–C bonds are formed, and the density of these hydrophobic groups is proportional to the density of the initial –OH groups. Another advantage of this functionalization is that the alkane chain can contain other functional groups (*e.g.* amino groups from γ-aminopropyl trimethoxysilane,[31] or alcohol groups from 3-aminopropyl-triethoxysilane[32]), which can be used for further polymer grafting. The disadvantage of this method is the cost and the strict high water-content requirement.

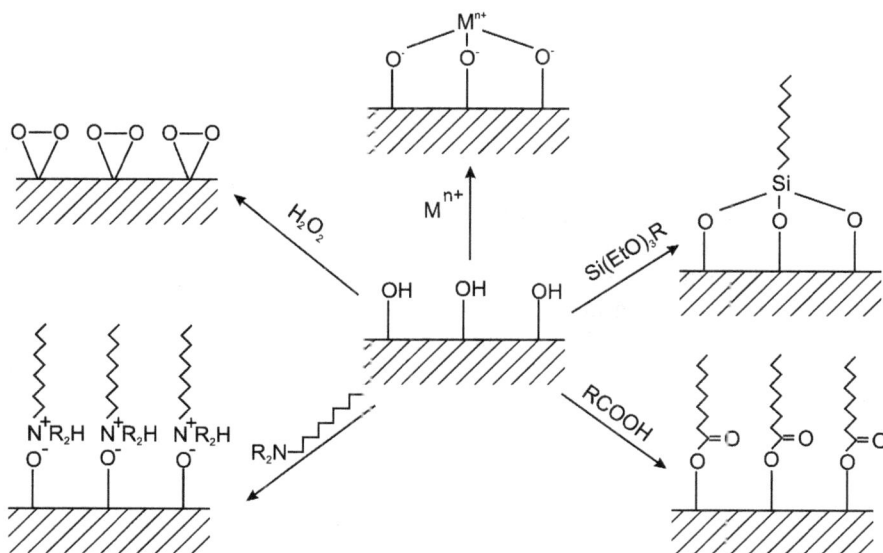

Figure 4.9 Examples of surface functionalization of titanate nanotubes using chemical reactions.

Another method of titanate nanotube surface hydrophobization is based on Van-der-Waals interactions between titanate nanotubes and cationic surfactant in aqueous solutions. The treatment of nanotubes with cetyltrimethylammonium bromide (CTAB)[33] or poly(diallyldimethylammonium) chloride[34] greatly increases the hydrophobic properties of nanotubes, due to a strong adsorption of surfactant on their surface. Moreover, the addition of surfactant in excess can switch over the zeta potential of nanotubes from negative to positive, due to the formation of admicelles.[33]

Although the surface –OH groups of titanate nanotubes are characterised by a weak acidity, it is possible to perform an esterification between them and various carboxylic acids[35] in anhydrous alcohol solvent. The surface of titanate nanotubes has been shown to bond relatively well with carboxylic acid, demonstrating monodentate, as well as bidentate, modes.

4.4   Stability of Nanotubes and Phase Transformations

In many applications, titanate nanostructures are exposed to chemically aggressive media, which can affect their stability. It is therefore important to understand the range of operational conditions under which nanotubes are stable, and the transformations to nanotubes which occur outside these conditions. Figure 4.10 shows transformations, occurring under various conditions, which result in a change in morphology or crystal structure of titanate nanotubes. All transformations can be grouped into three divisions, according to the type of treatment: thermal, chemical or mechanical.

4.4.1   Thermal Stability

A knowledge of the thermal stability of titanate nanotubes is important, since some applications or manipulations (including: catalyst supports or the curing of composite films) require an increased operating temperature. At elevated temperatures there are at least three processes occurring with protonated titanate nanotubes, namely: dehydration, crystal structure transformation and a modification in morphology. All three processes occur simultaneously, and each has a characteristic range of temperatures related to their particular phase transition. Figures 4.11a and b show typical TGA and DSC curves of protonated titanate nanotubes. The weight loss and exothermic processes that occur in the temperature range of 25 to 450 °C, are usually associated with the removal of water.

In the temperature range of 25 to 120 °C, desorption of water from nanotube pores results in a loss of approximately 7–8 wt%. A further increase in temperature up to 250 °C, results in the removal of crystallographic water from nanotubular titanates. This process is accompanied by a decrease in interlayer spacing d_{200} and an increase in density, due to the weight loss and shrinkage of the nanotube volume. According to TGA data, approximately 3–4 wt% of mass is loss during the dehydration of crystallographic water. At temperatures

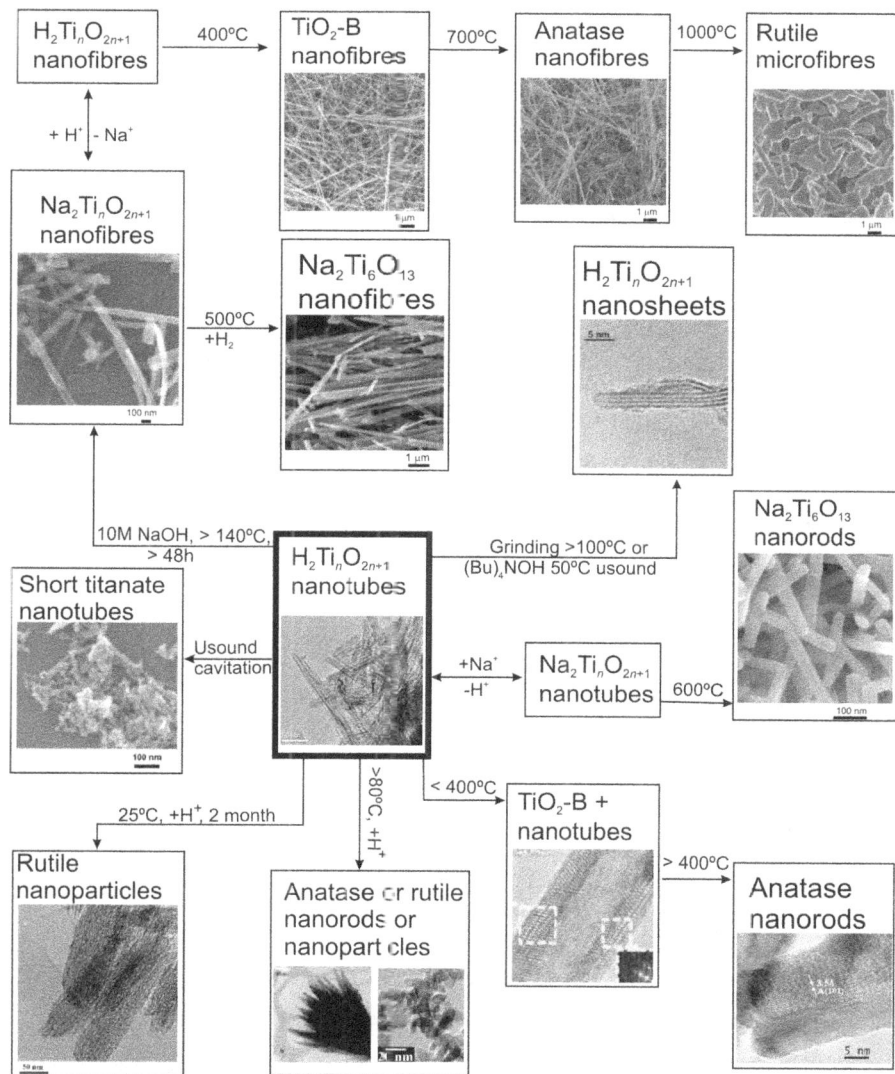

Figure 4.10 Chemical and structural transformations of titanate nanotubes and nanofibres. (The SEM and TEM images are reproduced from refs. 36,39,45,48,49,51,52).

above 250 °C, topotactic transformation of $H_2Ti_3O_7$ to the intermediate phases of $H_2Ti_6O_{13}$ and $H_2Ti_{12}O_{25}$ may occur,[36] resulting in a further decrease in interlayer spacing d_{200} and a further increase in density. Water loss accounts for an additional mass loss of approximately 1–2 wt%, according to TGA. At temperatures below 400 °C, the removal of the remaining water results in the formation of monoclinic TiO_2-(B) nanotubes (see Figure 4.11b). This monoclinic TiO_2-(B) phase can be detected using the characteristic reflection in the

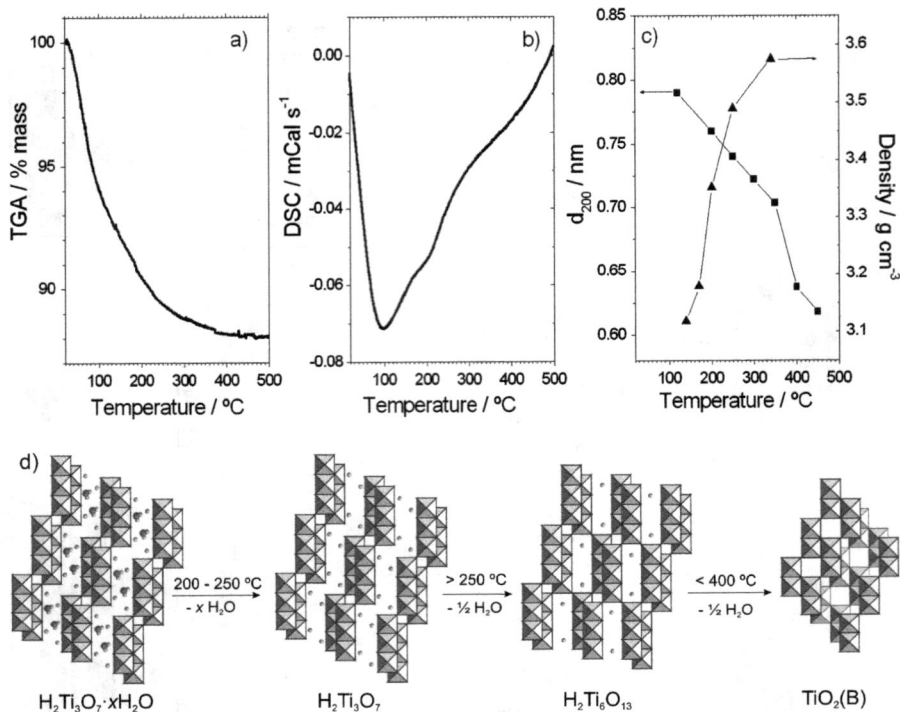

Figure 4.11   Thermal transformations of protonated titanate nanotubes. a) Graph of thermogravimetric analysis (TGA), b) differential scanning calorimetric (DSC) curve, c) graph showing density and interlayer distance d_{200} as a function of calcination temperature, and d) schematic crystal structures of hydrated trititanate and its transformation to hexatitanate, followed by the formation of TiO_2-(B). (Density was measured using helium adsorption, d_{200} structures adapted from ref. 36).

XRD pattern at *ca.* 15 °C (ref. 37,38). A further increase in temperature above 400 °C, results in the transformation of nanotubular TiO_2-(B) to anatase solid nanorods, accompanied by a filling of nanotubes hollow cavities and a loss of tubular morphology.

   The level of ion-exchanged sodium in nanotubular titanates can determine the thermal stability and the nature of thermal transformations during annealing.[39–42] When fully saturated with sodium ions, titanate nanotubes can lose their nanotubular morphology and convert to $Na_2Ti_6O_{13}$ nanorods only, at 600 °C.[43,44]

   Similar thermal transformations can occur in nanofibrous titanates (see Figure 4.10). Protonated nanofibres can under calcinations undergo the sequence transformations to nanofibrous TiO_2-(B) at 400 °C, followed by transformation to nanofibrous anatase at 700 °C, and then formation to microfibrous rutile at 1000 °C.[45,46] The formation of anatase nanofibres is also possible by the hydrothermal treatment of titanate nanofibres in water at 150 °C.[47] The sodium substituted nanofibres can be converted to $Na_2Ti_6O_{13}$ nanofibres at 500 °C in the presence of hydrogen.[48]

4.4.2 Acidic Environments

The phase transformation of titanate nanotubes at elevated temperatures during acid hydrothermal treatment, shows that both the anatase and rutile polymorphs can be formed in the presence of nitric acid ($1 \, mol \, dm^{-3}$), at temperatures above 80 °C after 48 hours of treatment.[49] The resulting anatase and rutile polymorphs have nanowire or nanocrystalline morphology. Acidic hydrothermal treatment of titanate nanotubes at 175 °C, results in the formation of polycrystalline anatase nanorods.[50] Titanate nanotubes have poor stability in diluted inorganic acids, even at room temperatures, and slowly transform to rutile nanoparticles over several months. The rate of this transformation depends on the nature of the inorganic acid, and is correlated to the solubility of titanates in the acid.[51]

Titanate nanotubes in their protonated form can be transformed to nanostructured anatase under hydrothermal conditions, even in the absence of acid addition.[52,53] This can be attributed to either the release of acid impurities left after the previous protonation of nanotubes, or the lower stability of protonated nanotubes.

It is possible that the acid-induced transformation of nanotubes to TiO_2 nanostructures occurs by dissolution of the initial nanotubes, accompanied by the simultaneous crystallisation of TiO_2 nanostructures, since the rate of the transformation is proportional to the steady-state concentration of dissolved Ti(IV).[51]

4.4.3 Mechanical Treatment

Titanate nanotubes and nanofibres are fragile structures which can be readily fragmented by ultrasonic treatment of an aqueous suspension, resulting in a decrease in the average length of nanotubes or a splitting and shortening of the nanofibres.[54] This can provide a convenient tool for controlling the nanotube aspect ratio allowing, for example, a reduction in the average length of nanotubes to 50 nm after 3 hours treatment in a conventional ultrasonic bath.

It has recently been shown that the process of nanosheet scrolling into nanotubes can be reversed by mechanical treatment at elevated temperatures. For example, the grinding of protonated nanotubes at 100 °C for 45 minutes, results in the disappearance of nanotubes and the formation of small multi-layered nanosheets.[55] Short-term ultrasonic treatment of titanate nanotubes in the presence of tetrabutylammonium hydroxide (TBAOH), followed by soaking for 3–7 days at 50 °C, also results in an unwrapping of nanotubes to form nanosheets.[56]

References

1. P. W. Voorhees, *J. Stat. Phys.*, 1985, **38**(1–2), 231.
2. R. Digilov, *Langmuir*, 2000, **16**, 1424.
3. M. Alcoutlabi and G. B. McKenna, *J. Phys.: Condens. Matter.*, 2005, **17**, R461.
4. R. Ma, T. Sasaki and Y. Bando, *Chem. Commun.*, 2005, **7**, 948.
5. D. V. Bavykin, A. A. Lapkin, P. K. Plucinski, J. M. Friedrich and F. C. Walsh, *J. Catal.*, 2005, **235**, 10.
6. X. Sun and Y. Li, *Chem. Eur. J.*, 2003, **9**, 2229.
7. D. V. Bavykin, V. N. Parmon, A. A. Lapkin and F. C. Walsh, *J. Mater. Chem.*, 2004, **14**, 3370.
8. D. V. Bavykin and F. C. Walsh, *J. Phys. Chem. C*, 2007, **111**, 14644.
9. T. Dickenson and A. Fiennes, *Chemical Kinetics*, Pergamon, 1966.
10. J. Crank, *The Mathematics of Diffusion*, Clarendon Press, Oxford, 1956.
11. J. Mon, Y. Deng, M. Flury and J. B. Harsh, *Microporous Mesoporous Mater.*, 2005, **86**, 277.
12. I. A. Stenina, A. D. Aliev, P. K. Dorhout and A. B . Yaroslavtsev, *Inorg. Chem.*, 2004, **43**, 7141.
13. D. R. Lide, *CRC Handbook of Chemistry and Physics*, CRC press, 84th edn, 2003.
14. D. V. Bavykin, A. A. Lapkin, P. K. Plucinski, L. Torrente-Murciano, J. M. Friedrich and F. C. Walsh, *Top. Catal.*, 2006, **39**(3–4), 151.
15. F. C. Walsh, D. V. Bavykin, L. Torrente-Murciano, A. A. Lapkin and B. A. Cressey, *Trans. Inst. Met. Finish*, 2006, **84**, 293.
16. M. Hodos, E. Horvath, H. Haspel, A. Kukovecz, Z. Konya and I. Kiricsi, *Chem. Phys. Lett.*, 2004, **399**, 512.
17. H. Li, B. Zhu, Y. Feng, S. Wang, S. Zhang and W. Huang, *J. Solid State Chem.*, 2007, **180**, 2136.
18. Q. Cheng, V. Pavlinek, Y. He, C. Li, A. Lengalova and P. Saha, *Eur. Polymer J.*, 2007, **43**, 3780.
19. H. G. Yang and H. C. Zeng, *J. Am. Chem. Soc.*, 2005, **127**, 270.
20. S. Uchida, R. Chiba, M. Tomiha, N. Masaki and M. Shirai, *Electrochem.*, 2002, **70**, 418.
21. J. Li, Z. Tang and Z. Zhang, *Electrochem. Commun.*, 2005, **7**, 62.
22. M. Miyauchi and H. Tokudome, *Appl. Phys. Lett.*, 2007, **91**, 043111.
23. M. Kitano, R. Mitsui, D. R. Eddy, Z. M. A. El-Bahy, M. Matsuoka, M. Ueshima and M. Anpo, *Catal. Lett.*, 2007, **119**, 217.
24. M. Miyauchi and and H. Tokudome, *J. Mater. Chem.*, 2007, **17**, 2095.
25. G. Cao, *J Phys. Chem. B*, 2004, **108**(52), 19921.
26. J. Yu and M. Zhou, *Nanotechnology*, 2008, **19**, 045606.
27. G. S. Kim, S. G. Ansari, H. K. Seo, Y. S. Kim and H. S. Shin, *J. Appl. Phys.*, 2007, **101**, 024314.
28. N. Wang, H. Lin, J. Li, X. Yang and B. Chi, *Thin Solid Films*, 2006, **496**, 649.

29. M. Takahashi, Y. Okada and K. Kobayashi, *Chem. Lett.*, 2008, **37**(3), 276.
30. M. T. Byrne, J. E. McCarthy, M. Bent, R. Blake, Y. K. Gunko, E. Horvath, Z. Konya, A. Kukovecz, I. Kiricsi and J. N. Coleman, *J. Mater. Chem.*, 2007, **17**, 2351.
31. H. J. Song, Z. Z. Zhang and X. H. Men, *Eur. Polym. J.*, 2008, **44**, 1012.
32. Z. Shi, G. Xueping, S. Deying, Y. Zhou and D. Yan, *Polymer*, 2007, **48**, 7516.
33. H. Niu, Y. Cai, Y. Shi, F. Wei, S. Mou and G. Jiang, *J. Chromatogr., A*, 2007, **1172**, 113.
34. R. Ma, T. Sasaki and Y. Bando, *J. Am. Chem. Soc.*, 2004, **126**, 10382.
35. W. Wang, J. Zhang, H. Huang, Z. Wu and Z. Zhang, *Colloids Surf. A*, 2008, **317**, 270.
36. E. Morgado Jr, P. M. Jardim, B. A. Marinkovic, F. C. Rizzo, M. A. S. Abreu, J. L. Zotin and A. S Araujo, *Nanotechnology*, 2007, **18**, 495710.
37. D. V. Bavykin, J. M. Friedrich and F. C. Walsh, *Adv. Mater.*, 2006, **18**, 2807.
38. G. Armstrong, A. R. Armstrong, J. Canales and P. G. Bruce, *Chem. Commun.*, 2005, **19**, 2454.
39. M. Qamar, C. R. Yoon, H. J. Oh, D. H. Kim, J. H. Jho, K. S. Lee, W. J. Lee, H. G. Lee and S. J. Kim, *Nanotechnology*, 2006, **17**, 5922.
40. E. Morgado Jr, M. A. S. Abreu, O. R. C. Pravia, B. A. Marinkovic, P. M. Jardim, F. C. Rizzo and A. S. Araujo, *Solid State Sci.*, 2006, **8**, 888.
41. E. Morgado Jr, M. A. S. Abreu, G. T. Moure, B. A. Marinkovic, P. M. Jardim and A. S. Araujo, *Mater. Res. Bull.*, 2007, **42**, 1748.
42. C. K. Lee, C. C. Wang, M. D. Lyu, L. C. Juang, S. S. Liu and S. H. Hung, *J. Colloid Interface Sci.*, 2007, **316**, 562.
43. X. Sun and Y. Li, *Chem. Eur. J.*, 2003, **9**, 2229.
44. O. P. Ferreira, A. G. Souza-Filho, J. Mendes-Filho and O. L. Alves, *J. Braz. Chem. Soc.*, 2006, **17**(2), 393.
45. S. Pavasupree, Y. Suzuki, S. Yoshikawa and R. Kawahata, *J. Solid State Chem.*, 2005, **178**, 3110.
46. R. Yoshida, Y. Suzuki and S. Yoshikawa, *J. Solid State Chem.*, 2005, **178**, 2179.
47. H. Yu, J. Yu, B. Cheng and M. Zhou, *J. Solid State Chem.*, 2006, **179**, 349.
48. Y. V. Kolen'ko, K. A. Kovnir, A. I. Gavrilov, A. V. Garshev, J. Frantti, O. I. Lebedev, B. R. Churagulov, G. V. Tendeloo and M. Yoshimura, *J. Phys. Chem., B*, 2006, **110**, 4030.
49. H. Y. Zhu, Y. Lan, X. P. Gao, S. P. Ringer, Z. F. Zheng, D. Y. Song and J. C. Zhao, *J. Am. Chem. Soc.*, 2005, **127**, 6730.
50. J. N. Nian and H. Teng, *J. Phys. Chem., B*, 2006, **110**, 4193.
51. D. V. Bavykin, J. M. Friedrich, A. A. Lapkin and F. C. Walsh, *Chem. Mater.*, 2006, **18**, 1124.
52. Y. Mao and S. S. Wong, *J. Am. Chem. Soc.*, 2006, **128**, 8217.

53. N. Wang, H. Lin, J. Li, L. Zhang, C. Lin and X. Li, *J. Am. Ceram. Soc.*, 2006, **89**(11), 3564.
54. D. V. Bavykin and F. C. Walsh, *J. Phys. Chem. C*, 2007, **111**, 14644.
55. G. Mogilevsky, Q. Chen, H. Kulkarni, A. Kleinhammes, W. M. Mullins and Y. Wu, *J. Phys. Chem. C*, 2008, **112**, 3239.
56. T. Gao, Q. Wu, H. Fjellvag and P. Norby, *J Phys. Chem. C*, 2008, **112**, 8548.

CHAPTER 5

Potential Applications

The elongated morphology, the high specific surface area and semiconductor electronic properties, render nanostructured titanates and TiO$_2$ promising materials for many applications. In this chapter, the most important of these, covering the fields of chemistry, physics, engineering, mechanics and medicine, are critically reviewed and gaps in our current knowledge are identified.

5.1 Energy Conversion and Storage

The future shortage of fossil fuels drives the search for alternative, renewable sources of energy. The nanostructured elongated titanates and TiO$_2$ materials have already established a niche in renewable energy related technologies,[1] including: cathodes for lithium batteries, catalyst and membrane components for fuel cells, supports for light-adsorbing centres in dye-sensitised solar cells, together with hydrogen sensing and storage systems.

5.1.1 Solar Cells

Elongated titanates and TiO$_2$ nanostructures have been examined for use as an electrode for dye-sensitized solar cells (DSSC). Figure 5.1a shows the principle of DSSC utilising nanotubular titanates. The photoexcited molecule of dye adsorbed onto the surface of nanotubes injects an electron to the semi-conductor, which then diffuses towards the electron sink. The oxidised form of the dye is reduced by iodide ions in the electrolyte solution, and the iodine released is further reduced on the platinum counter electrode.

The potential advantage of titanate nanotubes as an electrode for DSSC is realised by exploiting the phenomenon of improved adsorption of the positively charged dyes from aqueous solution onto the surface of negatively charged titanate nanotubes (compared to TiO$_2$NP).[2,3] This enables a compact mono-layer of dye to be deposited with a capacity of over 1000 molecules per

RSC Nanoscience & Nanotechnology No. 12
Titanate and Titania Nanotubes: Synthesis, Properties and Applications
By Dmitry V. Bavykin and Frank C. Walsh
© Dmitry V. Bavykin and Frank C. Walsh 2010
Published by the Royal Society of Chemistry, www.rsc.org

a)

b)

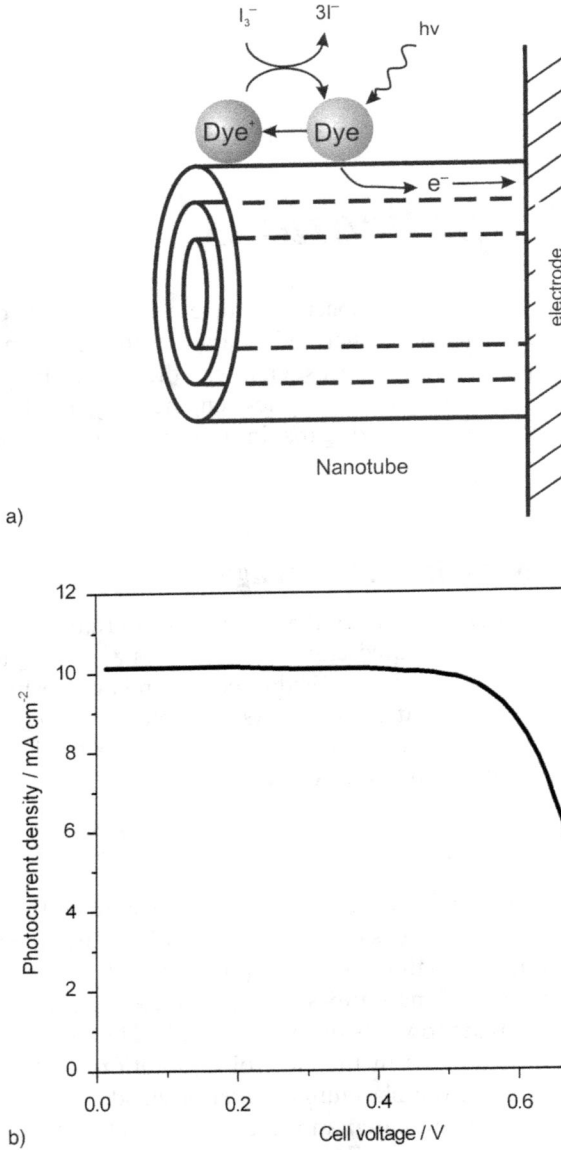

Figure 5.1   Dye-sensitised solar cells (DSSCs) which use titanate nanotubes as a support for the dye: a) the processes in DSSCs and b) a typical current–cell voltage curve for DSSC with TiO$_2$ NR electrode. (Images are adapted from ref. 10).

nanotube. Such a dense loading of dye allows the thickness of the light-adsorbing layer of the electrode to be significantly reduced from typical values of several microns, and the consequent reduction in the electron diffusion distance can potentially improve the charge collection. The second advantage of nanotubes is the elongated morphology of these semiconductors, which can provide a direct electron pathway from the point of injection to the electron sink, allowing improved electron transport and charge collection efficiency. This effect is particularly pronounced in TiO₂ nanotubular arrays.

Titanate nanotube-based electrodes

The design and manufacture of conventional DSSCs have been optimised for a nanoparticulate TiO₂ electrode, which can withstand high-temperature calcination and is a more suitable adsorbent for negatively-charged dye molecules, such as the *cis*-di(thiocyanate)bis(2,2-bipyridyl-4,4-di-carboxylate) ruthenium(II) complex. By contrast, titanate nanotubes are more suitable for the adsorption of positively-charged dyes[2] and are less stable during calcination at 450 °C (ref. 4,5), which is required for the conventional doctor blade method of electrode manufacture. Early data shows that titanate nanotubes can be readily applied as electrodes in a DSSC; however, no significant benefits were observed when compared with TiO₂ nanoparticle-based electrodes.[6] In most studies, however, calcination of the electrode at 450 °C was used to remove the polyethyleneglycol binder, meaning that TiO₂ anatase nanorods instead of H-TiNT were used in the DSSCs.

An alternative method for immobilising elongated titanates is the direct alkaline hydrothermal synthesis of nanotubes on the surface of titanium metal.[7,8] Prior to dye deposition at 500 °C, calcination of the sample was used, resulting in conversion of H-TiNT to TiO₂ anatase NR.[4] The advantage of the increased dye adsorption on the surface titanate nanotubes was not fully utilised.

In order to benefit from an improved electron transport in elongated nanostructures, it is necessary to assemble nanostructures on the surface of electrode. In most reports, elongated titanates are randomly oriented, which diminishes the advantage of direct transport. More recently, the successful alignment of titanate nanofibres on the surface of titanium under alkaline hydrothermal synthesis at 180 °C has resulted in an improved efficiency of DSSCs compared to the use of P25 TiO₂.[9]

Detailed studies of electron relaxation kinetics in elongated TiO₂ anatase nanorods have shown that the electron diffusion coefficient in nanorods is similar to that in P25. However, the electron lifetime was 3 times higher than that in P25, possibly due to a suppression of the recombination between electron and I⁻ ions on the surface of nanorods,[10] resulting in a higher value of the open circuit potential. Transient studies, using intensity-modulated photocurrent spectroscopy, show the improvement in charge collection in following order: P25, titanate nanotubes then anatase nanorods.[11] The charge collection ratio, which is defined as a ratio of the recombination time constant to the electron time collection (τ_r/τ_c), was found to be *ca.* 150, 50 and 10 for TiO₂ NR, H-TiNT and P25, respectively. These results were obtained for randomly oriented nanotubes and nanorods. The improvement in nanotube alignment on the electrode should offer further improvements in the electron collection efficiency.

The polycrystalline nanoparticles of anatase, obtained from titanate nanotubes by hydrothermal treatment at 240 °C in the presence of dilute HNO_3, have exhibited a slightly higher short circuit current, I_{sc}, but a lower open circuit potential, V_{oc} in DSSCs, compared to P25. This is most likely due to the higher recombination rate of charge carriers in these nanoparticles.[12]

The typical efficiency of a DSSC using elongated TiO_2 NR with standard dye and electrolyte is approximately 7.1% (ref. 10), and the shape of its voltammetric curve is shown in Figure 5.1 b. As a consequence of the leakage problems associated with a liquid electrolyte DSSC, recent attention has been focussed on solid or gel electrolyte DSSCs. The addition of H-TiNT filler to a polyethyleneglycol gel electrolyte up to a 10% of nanotubes has facilitated an improved ionic conductivity in gels.[13]

TiO₂ nanotube array electrodes

The other type of TiO_2 nanotubular array,[14,15] produced by anodising a Ti surface with a smaller specific surface area, also attracts attention as a candidate material for DSSC electrodes.[16] This interest is arises from the ordered structures within the electrode which allow an improvement in electron transport; the efficiency of this cell is claimed to be approximately 4.1%.

The classical Grätzel cell operates with sintered compressed layers of spheroidal TiO_2 NP as the electron harvesting material. A several micron thick agglomerate layer contains a high number of grain boundaries that act as recombination sites, reducing current collection efficiency. The ordered layer of aligned nanotubes can potentially show a lower rate of recombination due to the direct transport of electrons, but it is important to optimise the length of the nanotubes.

Figure 5.2 shows the dependence of photocurrent collected from the TiO_2 nanotube array electrode on the length of nanotubes. The nanotubes were functionalized with a Ru–complex deposited onto their surface and were illuminated at the 650 nm. It is evident that for thin layers, an increase in tube length results in a linear increase in the photocurrent, indicating the independence of current collection efficiency on nanotube length. When the thickness of the nanotube layer exceeds a certain value, a further increase in the nanotube length leads to a decrease in the photocurrent. Such a drop in photocurrent can be attributed to an increase in the electron diffusion length, resulting in an increase in the recombination rate. In the thick electrode layer, the absorption of light can occur unevenly causing a shadowing of areas having efficient electron transport.

The adsorption capacity of standard Ru–complex dye on the surface of anodic TiO_2 nanotubes is similar to that on the surface of TiO_2 anatase NP, resulting in relatively low coverage and making it necessary to use a thick layer of TiO_2 in order to absorb all incident light. For example,[17] the surface coverage of a 6 μm

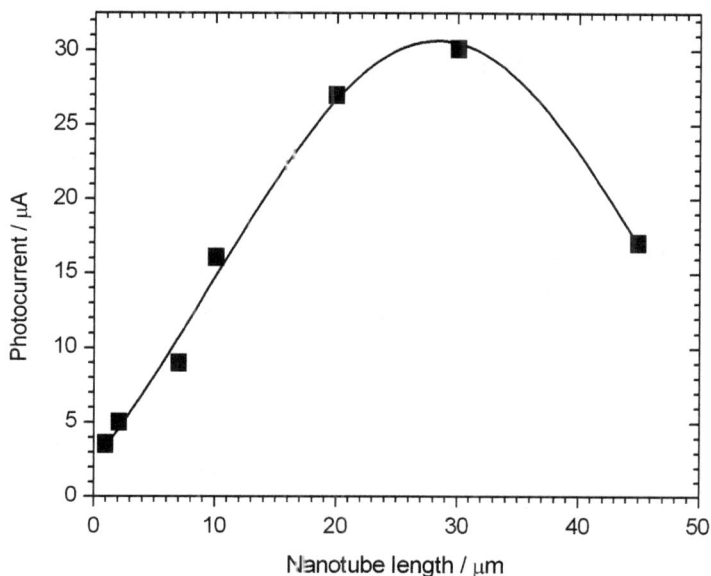

Figure 5.2 A graph showing the photocurrent collected from a Ru-complex dye-sensitised array of TiO_2 nanotubes as a function of nanotube length. Illumination at 650 nm. (Data adapted from ref. 15).

film of TiO_2 NP with standard N719 dye is *ca.* $0.5\,nmol\,cm^{-2}$, resulting in 80% absorption at 535 nm. In contrast, titanate nanotubes are characterised by a higher adsorption capacity, particularly towards negatively-charged dyes.[2]

The manufacture of DSSCs using TiO_2 nanotube arrays can be achieved either by the anodic oxidation of a titanium layer vacuum-deposited onto the surface of a conductive glass forming a transparent conductive glass/TiO_2 nanotube electrode, or by the anodic oxidation of titanium foil to form non-transparent electrodes, which are illuminated from the reverse side during operation.[14] The TiO_2 nanotubes obtained by template hydrolysis were also studied in a DSSC recently, with a reported efficiency of *ca.* 8.4%.[18]

It is useful to review the advantages and limitations of titanate and TiO_2 nanotubes when used in DSSC applications. These nanostructures can provide a high capacity for dye adsorption onto their surfaces, potentially allowing a reduction in the thickness of the light harvesting layer and a reduced electron diffusion time. However, it has proved difficult to prepare films of titanate nanotubes with a high degree of alignment, which has resulted in reduced current collection efficiency due to ineffective electron transport in the light harvesting layer. On the other hand, TiO_2 nanotube arrays are characterised by high nanotube alignment and relatively high current collection efficiencies. A low surface area of nanotubes and a poor dye adsorption capacity can result in an increased thickness of the light harvesting layer, which can lower the current collection efficiency.

5.1.2   Lithium Batteries

Nanostructured materials are widely used as electrodes in rechargeable lithium batteries.[19] Elongated titanates also attract attention as possible negative electrode material for lithium cells due to their open, meso-porous structure, efficient transport of lithium ions, and effective ion-exchange properties. Such properties result in these electrodes having a high charge/discharge capacity ($<300\,mA\,h\,g^{-1}$), and fast kinetics, together with very good robustness and effective safety characteristics.[20,21] These new electrodes can replace commercial, carbon-based negative electrodes, which suffer from safety concerns (due to the directional electrodeposition of lithium) and the formation of a solid–electrolyte interface (SEI) layer which leads to charge loss.

Figure 5.3 a shows the principle of lithium storage in nanotubular titanates. During charging, lithium ions from the electrolyte solution intercalate between layers in the wall along the axis of nanotubes,[22] followed by cathodic reduction to form the Li_xTiO_2 phase:

$$xLi^+ + TiO_2 + xe^- \rightleftarrows Li_xTiO_2 \qquad (5.1)$$

where x is the lithium insertion coefficient. The simplified product formula does not represent the crystal structure of TiO_2 or titanates, but it is convenient to express the value of x. The nature of Li_xTiO_2 intercalate is unclear (*e.g.* oxidative state of Ti and Li), but it is suggested that lithium atoms accommodate the positions of ion exchanged protons in titanate nanostructures or the channels in TiO_2-B nanofibres.[23,24] The charge capacity of lithium batteries depends on the availability of lithium reduction sites, and the power characteristics of the batteries are often dominated by the kinetics of lithium intercalation.

There are three consecutive steps in the process of charging/discharging the electrode, namely: (*i*) the diffusion of lithium ions in the electrolyte; (*ii*) the diffusion of intercalated ions/atoms; and (*iii*) the electrochemical reaction. Each of these stages, as well as electron transport, can be limiting in the overall process. The use of nanostructured titanates and TiO_2 significantly improves the rate of the diffusion of intercalated lithium ions due to their small size and the featured crystal structure of the material, which provides sufficient space (interlayer spacing in walls) for ionic transport.

At the present time, the following elongated nanostructures have been studied for lithium storage: titanate nanotubes[25–27] (TiNT), titanate nanofibres,[28–31] (TiNF), TiO_2-(B) nanotubes,[32,33] nanofibres[34–36] and TiO_2 (anatase) nanorods.[33,37] The last three nanostructures were obtained by calcination of the protonated forms of corresponded titanates (see Chapter 4, Figure 4.10). All of these structures demonstrate an impressive ability to store lithium ions (see Table 5.1).

Titanate nanotubes are characterised by a high initial specific capacity which can rapidly decrease from approximately 300 to *ca.* $180\,mA\,h\,g^{-1}$ within several

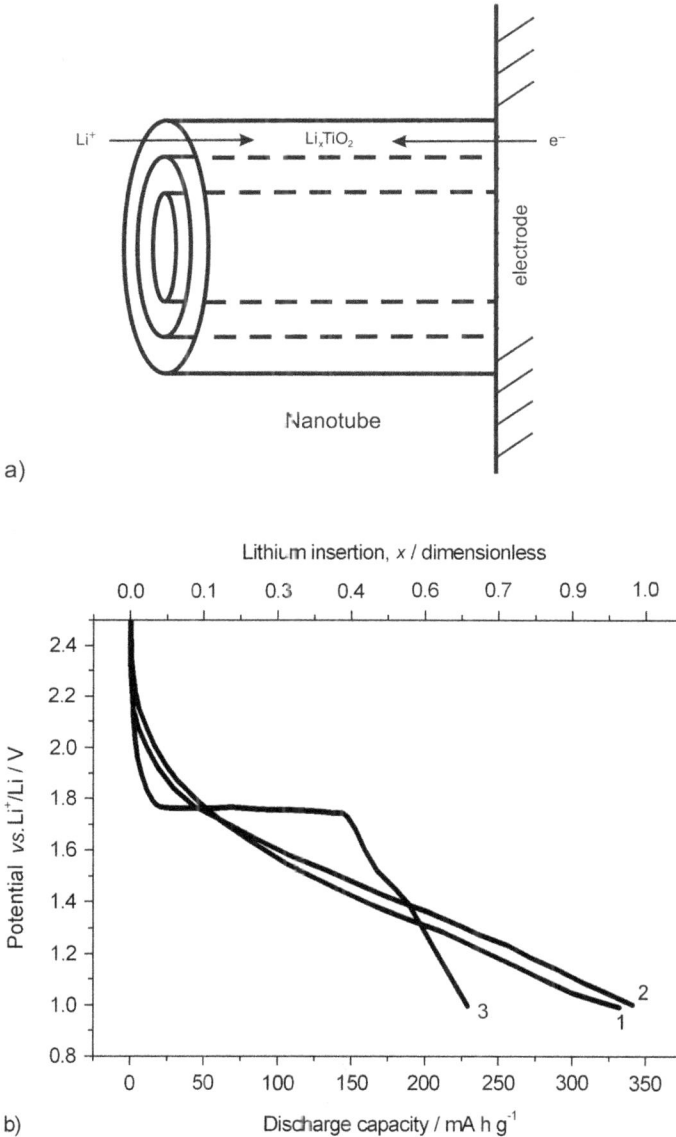

Figure 5.3 The principle of lithium storage in TiO_2 nanotubes: a) the principle and b) the first discharge curve for (1) TiNT, (2) TiO_2-(B) NT and (3) TiO_2 NR (adapted from ref. 33).

cycles. The charge/discharge curves are characterised by the absence of a plateau in which the voltage of the cell is constant. A slightly improved stability on cycling was observed using titanate nanofibres, which exhibited similar values for lithium storage capacity. Although these nanostructures have different

Table 5.1   Summary of electrodes containing elongated titanate or TiO$_2$ nanoparticles used for lithium storage. [P25: spheroidal TiO$_2$ nanoparticles (Degussa)].

Electrode	Discharge capacity^a / mA h g^{-1}	Specific current density / mA g^{-1}	Merits for lithium storage	Ref.
TiNT	220–250 170	110, 200 2000	Higher capacity than P25	25
TiNF	220 190 130	110 300 2500	Higher capacity than P25	28,29,30
TiO$_2$-(B) NT	240	50	Higher capacity than P25	33
TiO$_2$-(B) NF	200 100	200 2000	Higher capacity than P25	34,35
TiO$_2$ (anatase) NR	190 240	50 36	Plateau in current *vs.* potential curve	33,37
NiO/TiO$_2$-(B) NT	240 170	100 2000	Durability, lower electrical resistance	40
C–TiO$_2$ (anatase) NR	204	70	Lower resistance, plateau in current *vs.* potential curve	38
Co–TiNF	350	50	Intercalated Li affects magnetic properties	39
Co–TiO$_2$ (anatase) NF	140	50		
Ag/TiNT	190 160	50 600	Higher cycling stability at higher discharge rate	41
Sn/TiNT	312	30	SnLi alloying in pores of TiNT	42
TiO$_2$-(B) NT	296	25	Effect of electrode thickness on the discharge kinetics	43
TiO$_2$ (anatase) NR	215	25		

^{a}After 10 cycles.

morphologies and typical sizes, the rate of lithium ion intercalation is relatively fast in both as seen by their pseudocapacitive, faradaic behaviour.[26,30] This apparent inconsistency can be explained by taking into account the directions of lithium ion movement in nanotubes and nanofibres. It was suggested that in titanate nanotubes alkaline ions diffuse along their length, whereas in nanofibres they diffuse in a direction perpendicular to the length.[22] The typical length of nanotubes and width of nanofibres are several hundreds of nanometres, resulting in a characteristic diffusion time for intercalated lithium ion transport in both nanostructures of tens of minutes

The mechanism of lithium ion diffusion in TiO$_2$-(B) nanotubes and nanofibres is probably different to that in titanates; instead of diffusion between the layers of titanates, the lithium ions diffuse inside the smaller tunnels of theTiO$_2$-(B) crystals.[35] Due to an absence of ion-exchanged protons, hydrolysis of the electrolyte is suppressed. As a result, the initial discharge capacity does not deteriorate rapidly and the cycling stability is slightly better than that for nanostructured titanates. After a few cycles, the lithium storage capacity is similar to that of nanostructured titanates and the pseudocapacitive current behaviour also suggests that diffusion of lithium (external in electrolyte and internal intercalated) is not the limiting stage.

Anatase nanorods, produced by the calcination of TiNT, are characterised by a lower discharge capacity (see Table 5.1), a characteristic plateau on the charge/discharge curve (see Figure 5.3 b) and good reversibility. For all elongated nanostructures, the coulombic efficiency can approach 100%.

The power output demand for new generation lithium batteries (including power cells) is stimulating their use at higher current densities, which place additional requirements on the electrical conductivity of the nanostructured titanates and TiO_2 electrodes. Recent approaches to improve the conductivity of elongated structures include doping of TiO_2 nanorods with carbon;[38] doping TiNF and TiO_2NF with cobalt;[39] deposition of NiO particles on the surface of TiO_2-(B) nanotubes;[40] coating the surface of TiNT with silver nanoparticles;[41] and co-precipitation of a tin phase in the pores of TiNT. [42] All of these methods allow a reduced electrical resistance of the electrode, improve durability and decrease charge capacity degradation during high rate charge/discharge trials.

Another important issue for high current lithium batteries is the electrical resistance due to the limited mass transport of lithium in the electrolyte, resulting in a reduced voltage of the cell at high current densities for thick electrodes.[43] Potentially, this issue can be resolved by using 3-D electrodes of controlled hierarchy, which can be achieved, for example, by using TiO_2 nanotube arrays decorated with titanate nanotubes deposited inside the channels. The large pores would provide efficient and rapid transport of lithium ions from the cathode to the anode, while titanate nanotubes would supply sufficient sites for lithium intercalation.

Further improvements in lithium cell performance will probably focus on improvements in (*i*) the electrical conductivity of the elongated nanostructures;(*ii*) the mass transport of lithium ions in electrolyte by an optimal design of electrode porosity; (*iii*) the charge capacity of the electrode; and (*iv*) the stability of the cell with repeated cycling.

5.1.3 Fuel Cells and Batteries

Many developments in fuel cell technology have taken place over the last two decades as a result of concerns over the efficiency of and the environmental problems associated with existing energy converters. The nanostructured titanates and TiO_2 have been considered as supports for the electrocatalyst of fuel oxidation as shown in Figure 5.4 a. Early studies of palladium nanoparticles deposited on the surface of titanate nanotubes demonstrated the feasibility of nanostructured titanates for the oxidation of methanol in liquid fuel cells.[44] Further improvements in catalyst performance have been achieved by increasing the electrical conductivity of the elongated Pd/TiO_2 nanorods by

$BH_4^- + 8OH^- - 8e^-$

$BO_2^- + 6H_2O$ ◄─── Au

e⁻ ──►

electrode

Nanotube

a)

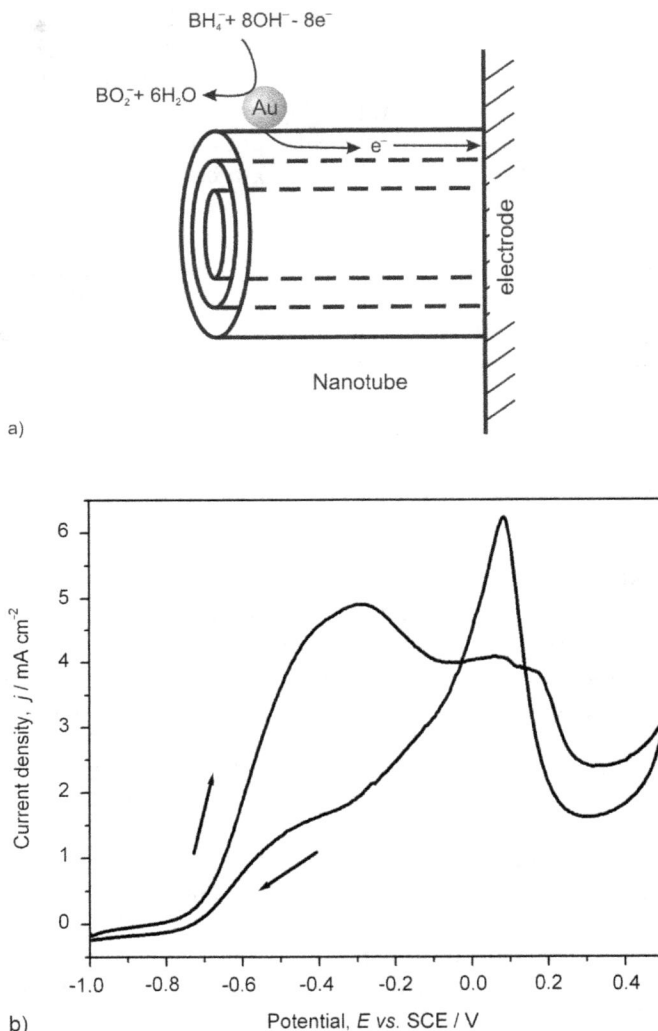

b)

Figure 5.4 Electron transport processes in a direct borohydride fuel cell: a) the principle and b) a cyclic voltammogram for the oxidation of borohydride ion using an Au/TiNT electrode (adapted from ref. 47).

carbon-coating *via* calcination of nanotubes coated with poly(ethylene glycol) at 600 °C.[45]

Recent studies have also showed that the simple addition of titanate nanotubes to the standard Pt/C catalyst, by mixing in slurry followed by drying on the electrode surface, results in an enhancement of catalyst performance. This is due to the structural water in the nanotubes and an increase in tolerance to CO poisoning, as a result of stimulating CO desorption from the catalyst surface.[46] Gold deposited onto titanate nanotubes has also demonstrated a performance comparable to commercial Au/C catalysts for the anodic oxidation of $NaBH_4$ in a direct borohydride fuel cell[47] (see Figure 5.4 b). The electrical charge during oxidation of the borohydride ion per unit mass of gold was approximately $8300\,mC\,cm^2\,mg^{-1}$ and $3900\,mC\,cm^2\,mg^{-1}$ for Au/TiNT and Au/C electrocatalysts, respectively (see Table 5.2).

Titanate nanotubes have been used not only for enhanced fuel cell catalyst efficiency, but also for improving proton conductive membrane performance. It has been shown that the addition of titanate nanotubes (up to 15 wt%) to the Nafion membrane, enhanced proton exchange conductivity at elevated temperatures (130 °C) due to water retention in the nanotubular titanates.[48] Such a composite membrane can be cast from a mixed slurry of nanotubes and 5% Nafion solution (DuPont), redissolved in dimethylsulfoxide.[48]

Applications using anodic TiO_2 nanotube array electrodes in fuel cell technology have not yet been realised, probably due to the relatively low surface area of nanotubes. In combination with other nanostructures, however, such an array can be used for structural reinforcement.

5.1.4 Hydrogen Storage and Sensing

Hydrogen storage has become a hot topic in recent years due to the possibility of the wide implementation of hydrogen as a major energy carrier in the hydrogen economy. Currently there are several commercially available technologies to store hydrogen, including: physiadsorption of hydrogen on high specific surface area materials at cryogenic temperatures, or chemisorption of hydrogen into nanoparticles of metal alloys forming hydrides.[49] In the first case, the interaction between hydrogen and adsorbent materials is relatively weak (*ca.* $10\,kJ\,mol^{-1}$; ref. 50), necessitating the use of low temperatures or a high pressure of the hydrogen in order to achieve high values of hydrogen uptake. In the case of chemisorption, the interaction between hydrogen and adsorbent materials is too strong (an order of magnitude of $100\,kJ\,mol^{-1}$; ref. 50), resulting in the need for high temperatures in order to release the gaseous hydrogen.

As a result of these limitations using known technologies, systems are currently being sought which provide a new type of interaction between hydrogen and hydrogen-capture materials, characterised by intermediate values of interaction energy (in a range of $30–60\,kJ\,mol^{-1}$). A bond between hydrogen and the adsorbent material of intermediate strength might facilitate a reversible sorption system, which can operate under near ambient conditions whilst allowing a high degree of hydrogen uptake.

The ability of titanate nanotubes to reversibly accumulate molecular hydrogen with a relatively high uptake and over a wide range of temperature (from −196 °C to 125 °C[51,52]), opens up the possibility of their use in hydrogen storage and related applications. Early data showed that values for the enthalpy and the activation energy of hydrogen sorption into titanate nanotubes could be estimated as *ca.* $−30\,kJ\,mol^{-1}$ and $44\,kJ\,mol^{-1}$, respectively.[51]

Table 5.2 Reported electrocatalytic processes using nanostructured titanates and TiO$_2$.

Catalyst Formula	Method of preparation	Particle size/nm	Loading	Process reaction	Activity or performance	Benefits	Ref.
Au/TiNT	Ion-exchange-reduction	4	0.1 mg cm^{-2}	BH$_4^-$ + 8 OH$^-$ $-8e^-$ → BO$_2^-$ + 6 H$_2$O	8320 mC mg^{-1} cm^{-2}	Borohydride fuel cell	47
Pd/TiNT	Ion-exchange-reduction	n/a	3 wt%	CH$_3$OH $- 6e^-$ + H$_2$O → CO$_2$ + 6H$^+$	n/a	Methanol fuel cell catalysts	44
Pd/TiNT	Ion-exchange-reduction	6–13	0.5 mg cm^{-2}	N$_2$H$_4$ $- 4e^-$ → N$_2$ + 4H$^+$	n/a	More active than Pd/TiO$_2$ NP	83
Pd/TiO$_2$ NR + C	Coprecipitation reduction	30	0.3 mg cm^{-2}	C$_2$H$_5$OH $- 12e^-$ + 3H$_2$O → 2CO$_2$ + 12H$^+$	n/a	TiO$_2$ + carbon composite improves conductivity	45
RuO$_2$/TiNT	Impregnation	n/a	n/a	CO$_2$ + 5H$_2$O + 6e$^-$ → CH$_3$OH + 6OH$^-$	Current yield = 60 %	Better performance compared to RuO$_2$/TiO$_2$	82
TiNT + Pt/C	Mixing in slurry	2	1 mg cm^{-2}	C$_2$H$_5$OH $- 12e^-$ + 3H$_2$O → 2CO$_2$ + 12H$^+$	120 mA mg^{-1} (Pt)	TiNT promotes Pt/C activity, tolerance to CO	46

TiNT: titanate nanotubes, TiNF: titanate nanofibres, NR: nanorods, NP: nanoparticles, n/a: not available.

Hydrogen uptake can exceed 2 wt% at 80 °C, but the characteristic time of sorption can be as slow as several hours.[51]

The exact mechanism and nature of bonded hydrogen are still under investigation. It has been suggested that hydrogen can occupy interstitial cavities between layers in the wall of nanotubes without chemical bond formation. The –OH groups in the nanotube lattice could stabilize the hydrogen molecules *via* weak van-der-Waals interactions. This could result in the formation of H-TiNT $\cdot xH_2$ clathrates, which might show similarities to the recently reported hydrogen clathrate hydrate, $(32 + x)H_2 \cdot 136H_2O$ (ref. 53). Such clathrate structure is possibly also produced in some nanotubular silicates, which, as was recently shown, can adsorb hydrogen up to 1.7 wt% at 25 °C (ref. 54). The ability of molecular hydrogen to form guest–host compounds with various host structures might be found in other porous, –OH-rich materials in the future.

Another interesting property of anodically produced TiO_2 nanotubes is that their electrical resistance dramatically decreases by up to seven orders of magnitude in the presence of sufficient hydrogen.[55] Systematic studies of this effect showed that longer nanotube arrays displayed a smaller drop in resistance, with significantly longer response/recovery times.[56] The nanotube wall thickness also affects the sensitivity of nanotubes to hydrogen. If the wall thickness is greater than an electron depletion region created by the chemisorption of oxygen, then the adsorption of hydrogen has little effect on the resistance of the sensor. By contrast, when the wall thickness is the same or less than the depletion region, the shift in electrical resistance of nanotubes under exposure to gaseous hydrogen can be very high. The highest sensitivity was achieved for nanotubes with a 13 nm wall thickness.[57] The presence of water and oxygen is found to decrease the sensitivity of the nanotubular arrays.

Titanate nanotubes[58] and Pt- or Pd-decorated titanate nanotubes[59] also possess a similar hydrogen-sensing behaviour, facilitating the manufacture of integrated systems, which can self control the amount of intercalated hydrogen. Nanofibrous TiO_2-(B) can also be used as a humidity sensor.[60]

5.2 Catalysis, Electrocatalysis and Photocatalysis

5.2.1 Reaction Catalysis

Elongated nanostructured titanates are characterised by a relatively high specific surface area, which is typically in the range of 200 to 300 $m^2 g^{-1}$ for nanotubes, and 20 to 50 $m^2 g^{-1}$ for nanofibres or nanorods. These values contrast with $<20 m^2 g^{-1}$ for TiO_2 nanotubular arrays produced by anodisation. The range of pore sizes (from 2 to 10 nm) ranks these materials as mesoporous; such structures are widely used in as a support in heterogeneous catalytic processes. The high surface area of the support facilitates high dispersion of the catalyst, while the open mesopores provide efficient transport of reagents and products.

Protonated titanate nanotubes can provide a moderate acid–base catalyst for esterification[61] and the hydrolysis of 2-chlorethyl ethylsulfide.[62] The Bronsted acidity of the nanotubular surface can also be increased by treatment with sulfuric acid, to give a value of approximately −12.7 on the Hammett scale.[61] Most of the catalytic studies of titanate nanotubes are, however, focussed on the utilization of its surface as a support for highly dispersed catalysts.

Several methods are used to deposit active catalysts into the pores of titanate nanotubes and nanofibres. Incipient wetness impregnation of the catalyst precursor, followed by thermal or chemical treatment is a common approach. This method allows the deposition of relatively large quantities of the catalyst; however, the dispersion and the distribution of catalyst are usually inferior to that obtained using other methods of deposition. A second method involves the precipitation of active materials on the surface of the nanotubes, which can be initiated by chemical, photo- or electro-chemical treatments. Such methods enable a better distribution of the catalyst, but the loading of metal is limited by the amount of precursor adsorbed onto the surface of the nanotubes. In addition, there is also the possibility of precursor precipitation in the bulk solution occurring as a side reaction.

Another *in situ* method for doping catalysts into titanate nanotubes is the addition of a catalyst precursor to the TiO_2–aqueous NaOH mixture prior to hydrothermal synthesis. The method allows catalyst atoms to be embedded into the crystal structure of the titanate nanotubes. A limitation of the method is that catalyst loading and dispersion are not easily controlled.

Since titanate nanotubes are characterised by good ion-exchange properties,[63] one method of catalyst deposition involves preliminary ion-exchange of the catalyst precursor in its cationic form with protons within nanotubular titanates (see Chapter 4, Figure 4.5). This allows an atomic-scale distribution of metal cations in the titanate lattice, achieving a higher metal loading compared to the adsorption of the precursor on the surface. Washing the sample with clean solvent, followed by reduction or chemical treatment, avoids precipitation of the catalyst in the bulk solution and leads to the formation of clusters or nanoparticles of catalysts evenly distributed on the nanotube surface only. A suitable choice of the ionic form of the metal precursor can significantly help to increase catalyst loading and to maintain high catalyst dispersion. For example, the use of a di-ethylendiamine complex of gold, $[Au(en)_2]^{3+}$, instead of tetra-chloroaurate, $[AuCl_4]^-$, increases the sorption of the gold precursor onto the titanate nanotubes by more than ten-fold.[64,65]

Examples of various metal catalysts deposited on the surface of titanate nanotubes are shown in Figure 5.5. The deposition method was ion-exchange followed by chemical treatment. Such an approach can achieve a relatively high catalyst loading, whilst maintaining a relatively small average particle size (see Table 5.3). Not all catalysts can be deposited using this method, for example, in cases where a suitable cationic form of the metal is unavailable in aqueous solution.

The most widely studied nanotubular supported catalyst is gold (Au/TiNT). The study of such materials parallels the work done on Au/TiO_2 catalysts, which are promising for low-temperature CO oxidation.[66,67] The early Au/TiNT catalysts demonstrate an activity comparable to that of the standard Au/

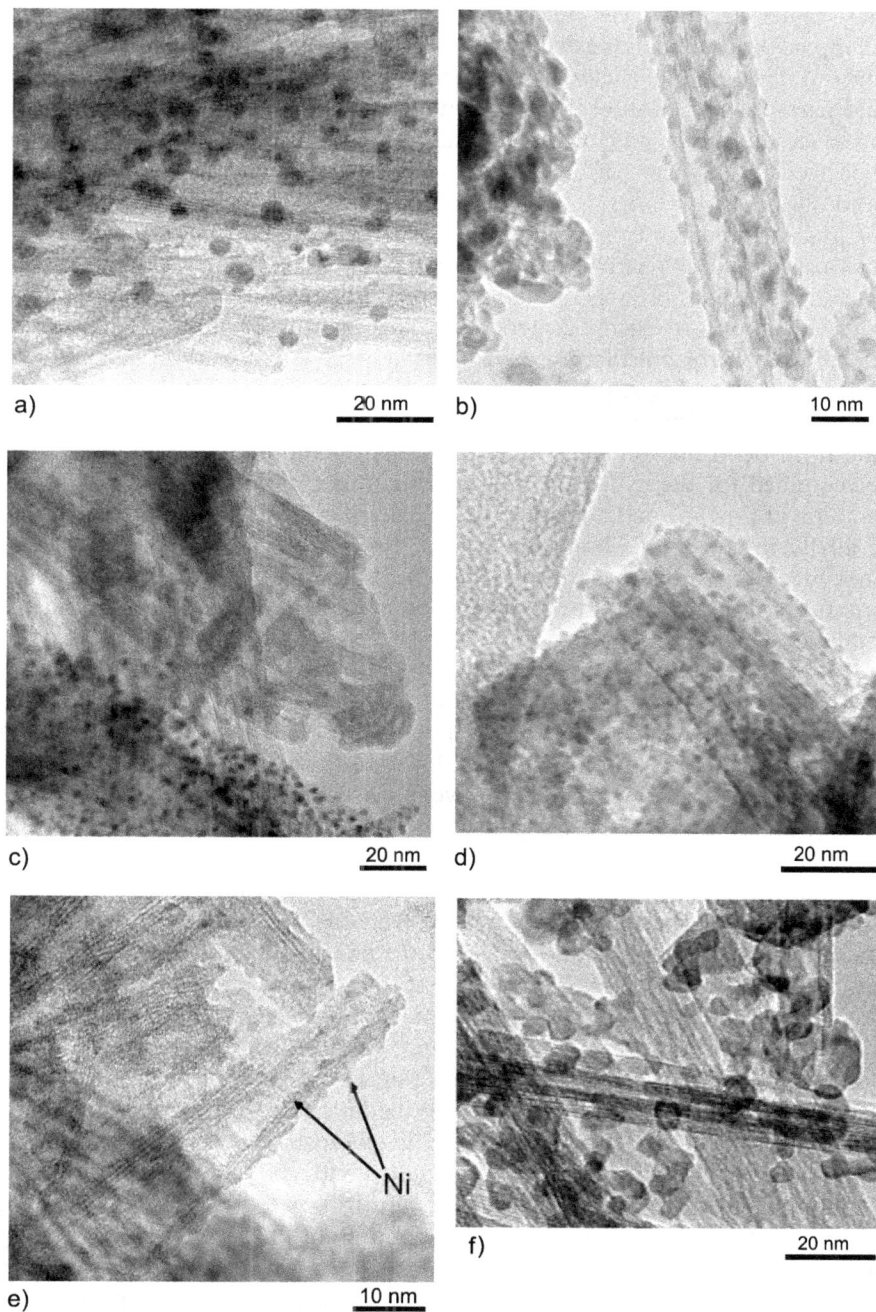

Figure 5.5 Examples of titanate nanotubes decorated with metal nanoparticles which are used in catalysis: a) Au/TiNT, b) Pd/TiNT, c) Pt/TiNT, d) RuOOH/ TiNT, e) Ni/TiNT and f) CdS/TiNT. (Images a), c) and d) are reproduced with kind permission from ref. 65 and image f) from ref. 119).

TiO$_2$ materials. However, recent improvements in performance have been attempted by the acid-assisted transformation of nanotubes (with deposited gold) to TiO$_2$ nanoparticles,[68] or by the high-temperature transformation of gold particles on nanotubes to those on nanorods.[69] Some successful attempts to reduce the use of precious metals such as gold have also been made using CuO (ref. 70) or Cu–Au[66] composites deposited on the surface of titanate nanotubes.

Catalysts prepared from gold and deposited on titanate nanotubes have also demonstrated a high activity for CO$_2$ reduction by hydrogen[71] and water shift reactions.[72] Most of these catalysts were prepared by precipitation from HAuCl$_4$ solution where the dispersity and loading of gold nanoparticles can be difficult to control. Further improvements in catalyst preparation utilising ion-exchange methods should lead to an improved catalyst activity.

The activity of palladium nanoparticle catalysts in a metal or metal hydroxide form deposited on the surface of titanate nanotubes or nanofibres, has been studied for the hydrogenation of phenol to cyclohexanone;[73] the hydrogenation of (*o*)-chloronitrobenzene to (*o*)-chloroaniline;[74] and the isomerisation of allylbenzene (double bond migration),[75] see Table 5.3. Copper(II) catalyst embedded into titanate nanotubes *in situ*, followed by conversion to TiO$_2$ nanotubes using calcination, showed good activity and high selectivity in the catalytic reduction of NO.[76]

Ruthenium(III) hydrated oxide (RuOOH) appears to be a promising catalyst for the selective oxidation of alcohols to aldehydes. The catalyst precursor is deposited onto the surface of the titanate nanotubes using ion-exchange with Ru^{3+} in aqueous solution, followed by hydroxylation with NaOH.[77] The resultant catalyst nanoparticles are evenly distributed on the surface of the titanate nanotubes, and the increase in RuOOH loading results in an increase in the catalyst particle density on the surface, rather than in the size of the particles (see Figure 5.6). Such conservation of the average particle size, and consequently the specific surface area, of RuOOH nanoparticles during changes in catalyst loading, is consistent with the independence of specific catalytic activity on loading. Figure 5.7 shows the specific catalytic activity (TOF) of the RuOOH/TiNT catalyst in the selective oxidation of benzyl alcohol to benzaldehyde as a function of catalyst loading.[77] Within the margins of error, the value of TOF remains unchanged over a range of loadings from 0.5 to 9 wt%.

Potential catalysts for the selective oxidation of dibenzothiophene by hydrogen peroxide at 60 °C are WO$_x$/TiO$_2$ nanostructures, obtained by the calcination of titanate nanotubes impregnated with (NH$_4$)$_2$WO$_4$ (ref. 78). This reaction models the process of the desulfurization of oil and the catalyst demonstrates a high activity. During preparation of the catalyst, an interesting morphological transformation has been reported.[79] The calcination of titanate nanotubes impregnated with (NH$_4$)$_2$WO$_4$ at 500 °C results in a collapse of the tubular structure and a release of the residue Na$^+$ ions to the surface of fibrous anatase. This leads to the formation of highly dispersed Na$_x$(WO$_4$) nanoparticles in which tungsten is in a tetrahedral coordination, providing a high activity for selective oxidation.

Table 5.3 Reported catalytic processes using nanostructured titanates and TiO_2.

Catalyst Formula	Method of preparation	Particle size/ nm	Loading	Process reaction	Activity or performance	Benefits	Ref.
Au/TiNT, Au-Cu/TiNT	Deposition - precipitation	4–17	0.1–2 wt%	$CO + 0.5\,O_2 \rightarrow CO_2$	$T_{50\%}(°C) = 47$	Competitive with commercial Au/TiO_2	66,67
Au/TiNT	Deposition - precipitation	10	1 wt%	$CO + 0.5\,O_2 \rightarrow CO_2$	$T_{50\%}(δC) = 123$	Conversion Au/TiNT to Au/TiO_2 NR	69
Au/TiO_2 NR	Adsorption -reduction	2–5	2–6 wt%	$CO + 0.5\,O_2 \rightarrow CO_2$	$T_{50\%}(°C) = 77$	Conversion Au/TiNT to Au/TiO_2 NP	68
Au/TiNT / Au/TiO_2 NP / Au/TiNT	Deposition - precipitation	3–5	1.5 wt%	$CO + H_2O \rightarrow CO_2 + H_2$	$T_{50\%}(°C) = 70$; $T_{50\%}(°C) = 25$; n/a	Early data show feasibility of catalyst	72
Cu(II)/TiO_2 NT	Impregnation or *in situ* deposition	n/a	2 wt%	$6\,NO + 4\,NH_3 \rightarrow 5\,N_2 + 6\,H_2O$	$T_{50\%}(°C) = 150$	High dispersion of catalyst	76
CuO/TiNT	Impregnation -calc. 200 °C	>5	6 wt%	$CO + 0.5\,O_2 \rightarrow CO_2$	$T_{50\%}(°C) = 90$	High activity	70
Pd/TiNT, Pd/TiNF	Impregnation - reduction	n/a	1 wt%	phenol + H_2 → cyclohexanone	$TOF \sim 94\,h^{-1}$, Select = 99%	High activity, selectivity and deactivation resistance	73
Pd/TiNT, Pd/TiNF	Impregnation - reduction	2–3	1–7 wt%	2-chloronitrobenzene + H_2 → 2-chloroaniline	$TOF = 186\,h^{-1}$, Select = 84 %	Selectivity can be controlled by Pd size	74
Pd(II)/TiNT	Ion-exchange	2–5	10 wt%	allylbenzene → (prop-1-enyl)benzene	$TOF = 2.3\,h^{-1}$, Sel. = 93 %	High dispersity of the catalyst at high loading	75
Pt/TiO_2 NT	*In situ* re-crystallisation	0.3– 3	30 wt%	$CO + 3\,H_2 \rightarrow CH_4 + H_2O$; $CO + H_2O \rightarrow CO_2 + H_2$	$Sel(CH_4) = 77\%$, $TOF = 20\,h^{-1}$	Higher performance than Pt/TiO_2	81
Pt/TiNT, Au/TiNT	Adsorption-photoreduction	2, 10	2 wt%	$CO_2 + 3\,H_2 \rightarrow CH_3OH + H_2O$	n/a	Early data shows feasibility of catalyst	71
Ru(III)/TiNT	Ion-exchange-alkali treatment	1– 2	1– 9 wt%	benzyl alcohol (–OH) → benzaldehyde (–CHO)	$TOF = 450\,h^{-1}$, Sel. = 99	High dispersity of catalyst at high loading	77
TiNT	hydrothermal	n/a	n/a	$C_2H_5SC_2H_4Cl + H_2O \rightarrow C_2H_5SC_2H_4OH + HCl$	n/a	Feasibility of titanate nanotubes in hydrolysis	62
WO_x/TiNT, WO_x/TiO_2 NF	Impregnation-calcined at 500 °C	<1	5–20	dibenzothiophene $\xrightarrow[60\,°C]{H_2O_2}$ dibenzothiophene sulfone	$TOF = 54\,h^{-1}$	Route to obtain highly dispersed WO_x on TiO_2	78

TiNT: titanate nanotubes, TiNF: titanate nanofibres, NR: nanorods, NP: nanoparticles, n/a: not available.

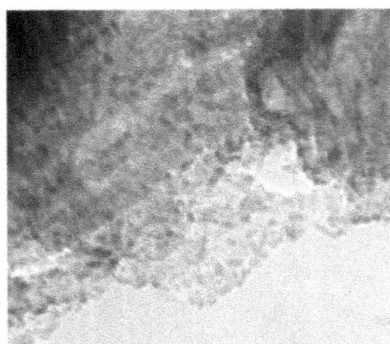
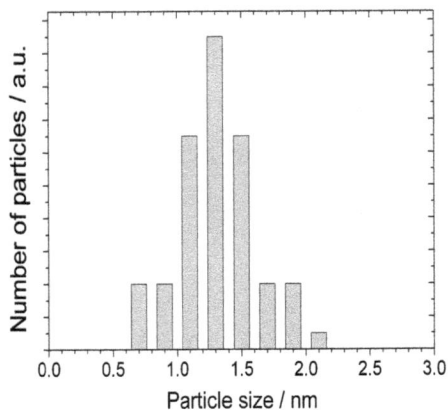

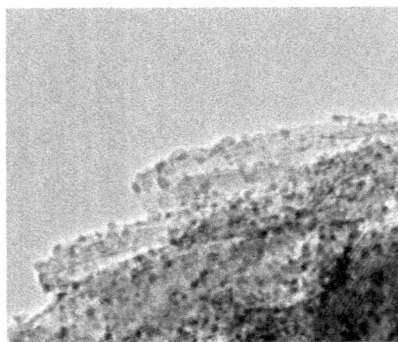
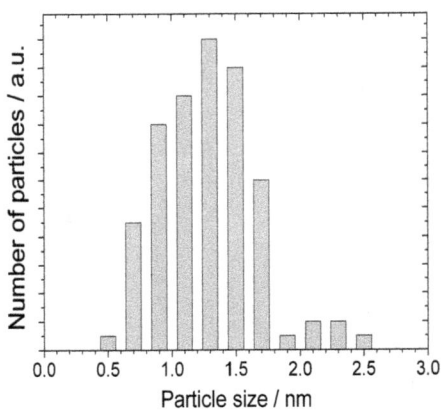

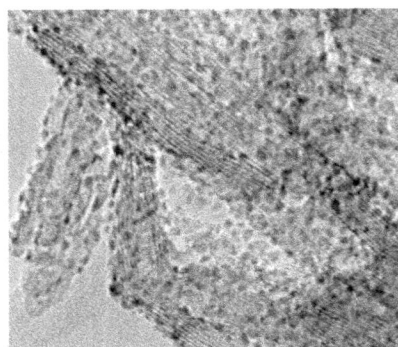
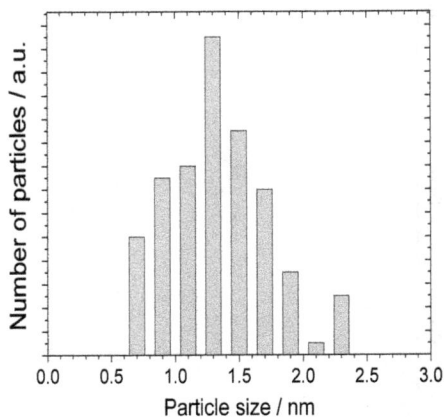

Figure 5.6 TEM images of ruthenium(III) hydrated oxide nanoparticles deposited onto TiO_2 nanotubes and histograms of particle-size distribution for metal loadings of: a) 1.1 wt.%, b) 3.4 wt.% and c) 8.7 wt.%. (Images are reproduced with kind permission from ref. 64).

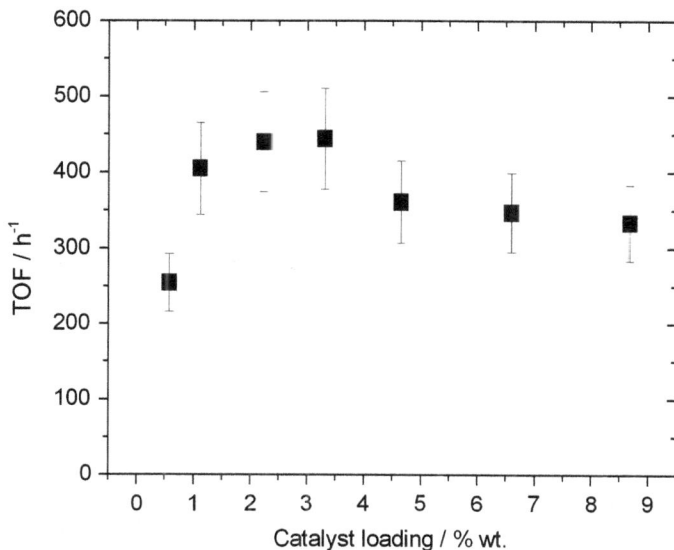

Figure 5.7   The specific catalytic activity of a RuOOH/TiNT catalyst as a function of RuOOH nanoparticle loading. The turnover frequency (TOF) is the rate of benzyl alcohol oxidation with oxygen in a continuous flow reactor at 117 °C. (Data are reproduced with kind permission from ref. 77).

Platinum deposited on the surface of wide-pore nanotubular catalyst supports can be synthesised using $[Pt(NH_3)_4](HCO_3)_2$ complex as a template.[80] In this method, the template is also a catalyst precursor. Such Pt/TiO_2 nanotubular catalysts can be characterized by their wide diameter (100–200 nm), and the Pt nanoparticles are well dispersed on the nanotubular surfaces with a high loading. Catalysts show a good performance in the CO reduction with H_2 selectively forming CH_4, as well as in the water shift reaction.[81]

All reported applications of titanate nanotubes in catalysis emphasise the benefits of a high surface area, together with the versatility of the surface chemistry and electronic interactions between the catalyst and the support, which allow an improved catalytic activity. The low cost of titanate nanotubes opens a novel route for nanostructured TiO_2-supported catalysts using either acid or thermal transformations. The control of the localisation of catalyst (either inside the hollow of the tubes or on the external surface) is still challenging[64] and represents an important topic for further studies.

5.2.2   Supercapacitors and General Electrochemistry

Composite electrodes consisting of titanate nanotubes supporting nanoparticles of precious metal have been also studied for electrocatalytic processes. These include the electrochemical reduction of CO_2 to methanol using a $RuO_2/$ TiNT composite electrode[82] and the oxidation of hydrazine (N_2H_4) using palladium-decorated titanate nanotubes.[83] In both cases, the composite electrodes demonstrated a better performance than TiO_2-based ones.

The oxides of transition metals, which can have several valence states, are possible materials for electrical energy storage in electrochemical capacitors. Effective dispersion of these oxides on the electrode surface is critical to yield a high capacitance, which can be achieved using titanate nanotubes as a support for the metal oxide. Early reports of vanadium(v) oxide deposited on the surface of titanate nanotubes showed the feasibility of such composites as capacitors.[84] Further improvements in capacitance stimulated the use of ruthenium (RuO_2)[85,86] or mixed ruthenium chromium oxides $(Ru_{1-y}Cr_yO_2$; ref. 87) deposited onto the surface of titanate nanotube composites, allowing values of specific (per mass of RuCr oxide) capacitance of *ca.* 1272 F g^{-1} to be obtained for a 4 wt% of $Ru_{1-y}Cr_yO_2$ loading (see Table 5.4). This value is almost twice that for the electrochemical capacitance of the bulk $Ru_{1-y}Cr_yO_2$ positive electrode. The high cost of ruthenium has encouraged the search for lower-cost substitute elements, resulting in the synthesis of cobalt hydroxide[88] and cobalt nickel double hydroxide deposited on TiNT, which have a specific (per mass of metal oxide) capacitance of approximately 1000 F g^{-1}.[89]

5.2.3 Photocatalysis in Elongated Titanates and TiO$_2$

During the last three decades, titanium dioxide has been comprehensively studied as a wide band gap photocatalyst for the oxidation of organic compounds.[90] The best TiO_2-based catalysts are usually characterised by a highly crystalline structure (which can reduce the recombination of photogenerated carriers); a high specific surface area (for acceleration of the interfacial reaction rate); and an abundance of surface –OH groups (which are required for the generation of OH radicals during photocatalytic reactions). All of these features are intrinsic to elongated titanates and anodised arrays of TiO_2.

Photocatalysis in elongated titanates

The optical properties of titanate nanotubes have been recently studied by various methods. The absorption threshold determined from diffuse reflectance spectra is usually very close to the band gap of titanium dioxide,[91] which is *ca.* 3.2 eV. However, more accurate studies of diluted colloidal solutions of TiNT, which allowed the errors caused by elastic light scattering to be avoided, estimated the nanotube band gap as *ca.* 3.87 eV (ref. 92). This indicates that the band gap of titanate nanotubes is wider than that of TiO_2 and closer to that of titanate nanosheets (3.84 eV; ref. 93). Photoluminescence measured at $-196\,°C$ from powder-state samples usually shows a band at 2.4 eV (ref. 94), whereas the spectrum from nanotubes dispersed in water[92] shows a multiple-line spectrum with several characteristic bands at 3.99, 3.77, 3.54, 3.09, 2.94, 2.51, 2.38, 2.16, 2.08 and 1.99 eV. Discrepancies between these two methods are probably due

Table 5.4 Reported application of nanostructured titanates and TiO_2 as supercapacitor.

Catalyst Formula	Method of preparation	Particle size/ nm	Loading	Process reaction	Activity or performance	Benefits	Ref.
RuO_2/TiNT	Precipitation	n/a	4–21 wt%	$RuO_2 + H_2O + e^- \rightarrow RuOOH + OH^-$	$230\,F\,g^{-1}$	Early reports	85,86
VO_x/TiNF	Adsorption	n/a	n/a	$VO_2^+ + 2H^+ + e^- \rightarrow VO^{2+} + H_2O$	n/a	Capacitance is higher than for bulk V_2O_5	84
$Co(OH)_2$/ TiNT	Precipitation	n/a	25–75 wt%	$Co(OH)_2 + OH^- \rightarrow CoOOH + H_2O + e^-$	$229\,F\,g^{-1}$	Capacitance is higher than for bulk $Co(OH)_2$	88
$Ru_{1-y}Cr_yO_2$/ TiNT	Co-precipitation	n/a	4–23 wt%	$Ru_{1-y}Cr_yO_2 + H_2O + e^- \rightarrow Ru_{1-y}Cr_yOOH + OH^-$	$250\,F\,g^{-1}$	Lower cost than pure RuO_2	87
$Co(OH)_2 + Ni(OH)_2$/ TiNT	Co-precipitation	n/a	60 wt%	$(CoNi)(OH)_2 + OH^- \rightarrow (CoNi)OOH + H_2O + e^-$	$631\,F\,g^{-1}$	High capacitance	89

to the strong scattering of light from the solid powder samples, which can mask the photoluminescence signal.

Absorption of light by titanate nanotubes results in the generation of charge carriers, which can eventually relax into a single electron trapped oxygen vacancy (SETOV)[95] or be trapped by Ti^{4+} ions forming Ti^{3+} centres, which can cause visible light absorption (see Chapter 3). Transient studies of photo-generated charged carriers in titanate and TiO_2-(B) nanotubes have revealed that the lifetime of trapped electrons is longer than that of TiO_2 nanoparticles, suggesting an improved charge separation due to the elongated morphology.[96]

In the presence of oxygen and organic molecules, the photogenerated carriers undergo relaxation processes following the routes shown in Figure 5.8a. Photogenerated electrons diffuse onto the surface of nanotubes and usually reduce oxygen molecules to form peroxo-species. Photogenerated holes also diffuse to the surface and react with surface −OH groups forming OH radicals, which further react with any organic molecules present.[90] Conventional TiO_2 photocatalysts are operated based on this principle.

It is well known that Na^+ impurities can significantly decrease the photo-catalytic activity of TiO_2 acting as a recombination centre.[90] The high level of sodium ions retained in titanate nanostructures after alkaline hydrothermal synthesis, can also significantly reduce the photocatalytic activity of titanates. This was recently confirmed by observation of the negative effect of sodium content on the photocatalytic activity of titanate nanotubes during of oxidation of dyes.[97,98] In contrast, the removal of sodium ions by the protonation of titanate nanotubes results in luminescence quenching,[91] indicating that the centres of radiative recombination are associated with sodium sites. This is consistent with observation of the negative effect of sodium ions on the pho-tocatalytic activity of TiNT.

The photocatalytic activity of prepared titanate nanotubes was found to be smaller (though not zero) than that of the standard P25 catalyst in the reaction

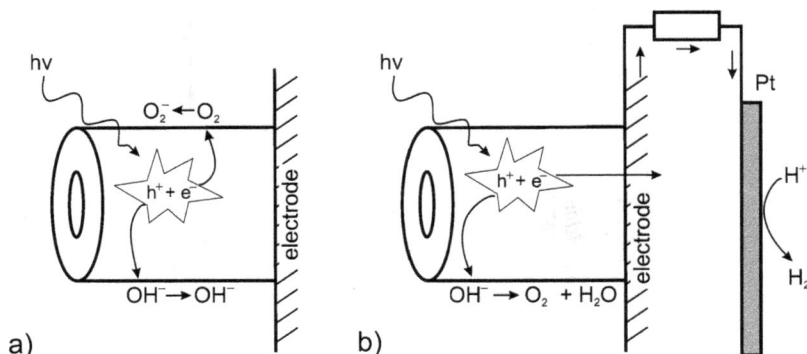

Figure 5.8   Photocatalytic processes during the oxidation of organics: a) initial pho-tocatalytic reactions and b) the process of photochemical water splitting on nanotubular TiO_2 and titanates.

of NH$_3$ oxidation,[99] as well as in the reaction of dye oxidation in aqueous suspensions.[100] This can be either attributed to impurities of sodium or to the moderate crystallinity of the 'as-prepared' titanate nanotubes. Further improvements in activity are focussed on the transformation of initial nanotubes to the more photocatalytically active forms of TiO$_2$.

Two methods for H-TiNT transformation have been reported, namely heat and heat/acid treatments. The anatase nanoparticles[101] or nanorods[102,103] produced by the calcination of H-TiNT at 400 °C, were characterised by an improved initial H-TiNT photocatalytic activity during the reaction of various organic molecules or dye oxidation. The increase in photocatalytic activity was accompanied by an improvement of nanostructure crystallinity. A further increase in calcination temperature resulted in a lower photocatalytic activity, due to stripping of the surface –OH groups and a reduction in the surface area of nanostructures. In contrast, the acid hydrothermal treatment of H-TiNT with residues of HCl at 200 °C (ref. 104), or H-TiNF with HNO$_3$ (0.1 mol dm^{-3}; pH 0–7) at 180 °C (ref. 105), resulted in the formation of nanostructured anatase with a fibrous or particulate morphology. This showed a good photocatalytic activity for the oxidation of model organic dyes. The method of acid hydrothermal transformation allows a reduction in synthesis temperatures and helps to keep surface –OH groups intact.

An additional improvement in charge separation in elongated titanates can be achieved using the recently discovered synergetic effect that takes place in mixed-phase nanocomposites. The effect occurs between two crystalline forms of TiO$_2$ (anatase and rutile) as a result of the small difference in flat band potentials, which stimulates a spatial separation of carriers thereby reducing their recombination.[106] The bi-crystalline mixture of TiO$_2$-(B) nanotubes–anatase nanoparticles (33 : 67%), prepared by calcination of H-TiNT, is characterised by an improved photocatalytic activity when compared with P25 TiO$_2$ for the hydrogen evolution from ethanol.[107] This approach has also demonstrated the versatility of this method for the easy preparation of mixed-phase nanocomposites. Titanate nanofibres[108] and nanotubes[109] decorated with TiO$_2$ (anatase) nanoparticles, deposited by hydrolysis of TiF$_4$ in the presence of H$_3$BO$_3$, are also characterised by good surface adhesion, increased surface area and improved photocatalytic activity.

Although TiO$_2$-based materials can be very active in photocatalytic reactions, their major drawback, delaying their widespread industrial use, is the relatively short wavelength of light necessary to participate in the photocatalytic reactions. The sensitisation of the photocatalyst to visible light is the "Holy Grail". Several approaches to the sensitisation of elongated titanate and TiO$_2$ nanostructures have been recently explored.

One of the successful ways to sensitise TiO$_2$ to visible light is to dope it with nitrogen, forming additional levels in the forbidden zone of wide band gap TiO$_2$. Several methods have been reported for doping titanate nanotubes with nitrogen. The first method involves the ion-exchange of ammonium (NH$_4^+$) ions with protons in H-TiNT, followed by calcination of the sample in air at 400 °C to form N-doped TiO$_2$- (B) nanotubes.[110,111] The second method employs the alkaline hydrothermal treatment of preliminary doped TiO_2 with nitrogen for the formation of N-doped titanate nanotubes.[112] The third method involves the nitration of TiNT with gaseous NH_3 at temperatures of 400–700 °C[113] at atmospheric pressure. This last method provides control over the level of N-doping to a nitrogen surface concentration of up to 12%. All N-doped elongated nanostructures show good photocatalytic activity under the visible range of incident light. The nature of photoactive centres in N-doped elongated titanates and TiO_2 is being actively studied. Recent XPS and ESR data show that nitrogen is most likely to be present in the form NO, occupying the interstitial positions between oxygen vacancies and Ti^{4+}-forming visible light absorption centres, and providing energy levels positioned within TiO_2 band gap. The photocatalytic activity in the visible range correlates with a concentration of these centres which is higher for N-doped anatase nanorods obtained by nitration of H-TiNT, than N-doped P25 prepared at 600 °C.[113]

Another method to sensitise TiO_2 to visible light is implantation with Cr(III) ions,[114] where Cr^{3+} ions occupy the positions of Ti^{4+} in the lattice and form electron levels in the forbidden zone. Photocatalytic activity under visible light is observed for catalysts with isolated Cr^{3+} ions inside the TiO_2 lattice. The agglomeration of chromium ions at higher chromium loading results in the appearance of recombination centres and a decrease in activity. There are several reports of chromium doping of titanate nanotube photocatalysts. Low levels of chromium doping (0.5 wt%) in titanate nanotubes can be achieved by the alkaline hydrothermal treatment of preliminary Cr-doped anatase.[115] The photocatalyst showed some activity under visible light during dye oxidation. The acid-assisted hydrothermal transformation of H-TiNT to anatase, in the presence of HNO_3 ($0.1 \, mol \, dm^{-3}$) and chromium(III) at 240 °C over 24 h, results in the formation of Cr-doped anatase nanoparticles.[116] The level of chromium in these Cr-doped anatase nanoparticles can be varied up to 10 wt%. However, the most active catalyst for photoelectrochemical water splitting was found to be a catalyst having a loading, of 3 wt%, in which no Cr_2O_3 phase formation was observed.

Several attempts to decorate titanate nanotubes with the narrow band gap semiconductor nanoparticles CdS [117–119] and ZnS[120] by ion-exchange, followed by H_2S treatment, have also been reported. Such heterojunction photocatalysts showed a moderate activity for dye oxidation, but the photocorrosion of sulfides is a major problem in such systems. Sensitisation of titanate nanofibres with NiO nanoparticles[121] or tin porphoryrin (Sn–TTP) complexes intercalated between layers of titanates[122] also resulted in photocatalytic activity of the composite material in the visible range. Femtosecond studies of the latter showed an effective charge separation of visible light photoinduced carriers. The photocatalytic oxidation of methyl orange showed a synergistic enhancement of activity using both UV and visible illumination.[122]

In conclusion, elongated titanate and TiO_2 nanostructures have been considered for photocatalytic processes, including: the oxidation of organic wastes in both air and water;[101,102,123] the splitting of water;[116,124] and the generation of hydrogen using sacrificial hole scavengers[107,125,126] (see Table 5.5). A comparison of elongated morphologies[125] for hydrogen evolution from methanol indicated that photocatalytic activity followed the trend: anatase NF > TiO_2-(B) NF ≫ H-TiNF. More extensive and systematic studies of all elongated morphologies, including: nanotubes, nanosheets and nanofibres transformed from each other by calcination or hydrothermal treatments, are needed.

Glass surfaces coated with photocatalytically active TiO_2 are already being commercially used as self-cleaning surfaces, due to their anti-fogging and super hydrophilic properties under UV light.[90] The elongated H-TiNT and TiO_2 anatase nanorods[127,128] demonstrate an even better surface wettability, as their tubular morphology results in an increased surface roughness, which can be beneficial in achieving a smaller contact angle between the surface coating and water droplets. A similar effect is also observed on anodic TiO_2 nanotube array surfaces, attributed to both the porous structure of the nanotubes and their high photocatalytic activity in the oxidation of hydrophobic molecules on the surface.[129]

Photochemical Water-splitting and Photocatalytic Oxidation on TiO₂ Nanotube Arrays

Anodic TiO_2 nanotube arrays have recently been thoroughly studied as a promising electrode for the photoelectrolysis of water and are reviewed elsewhere.[130] The principle of photochemical water-splitting is illustrated in Figure 5.8b. Absorption of light in nanotubular TiO_2 results in the generation of the main charge carriers: electrons and holes. Photogenerated holes migrate to the surface of the nanotubes and oxidise the water molecules, generating gaseous oxygen, whereas photogenerated electrons are collected to the external circuit and reduce the protons on the platinum counter electrode, generating gaseous hydrogen.

Nanotubular TiO_2 electrodes demonstrate an enhanced activity in the photoelectrolysis of water.[131] Such an enhancement can be attributed to significant improvements in light absorption at wavelengths near the band edge (375–400 nm) by trapping the light using photonic principles and the tubular morphology of material.[130] As a result, the near band-edge light can pass through the layer of TiO_2 several times, allowing the layer thickness to be decreased whilst maintaining a high level of light absorption. According to Figure 5.9, the photolysis of water is characterised by approximately –0.8 V *vs.* Ag/AgCl open-circuit potentials, a short circuit current density of $> 10\,\text{mA cm}^{-2}$ and a photoconversion efficiency, η, of approximately 12–15%. The latter value can be estimated as follows:[130]

$$r = \frac{(1.229 - V_{bias})I_p}{P_t} \times 100 \qquad (5.2)$$

where 1.229 V is the standard cell voltage required for the electrolysis of water (the difference between the standard redox potentials of oxygen and hydrogen

Table 5.5 Reported photocatalytic processes using nanostructured titanates and TiO_2.

Catalyst Formula	Method of preparation	Particle size/ nm	Loading	Process reaction	Activity or performance	Benefits	Ref.
$Pt/(TiO_2)$-(B) NT–anatase NP	Calcination at 400 °C	10 nm anatase	1 wt% Pt 33% TiO_2-(B)	$CH_3CH_2OH + hv \rightarrow CH_3CHO + H_2$	20% higher than Pt/P25	Facile method to bi-crystalline catalyst	107
$Cr(III)/TiO_2$ NP	Hydrothermal acid of Cr–TiNT	5 nm	3 wt%	$H_2O + hv \rightarrow H_2 + O_2$	n/a	Novel route for ion implantation	116
TiO_2 NP (anatase)	Calcination H-TiNT		n/a	Photocatalytic oxidation of organics	Higher than P25	Novel route for anatase nanoparticles	101
TiO_2 NR (anatase)	Calcination H-TiNT	10 × 100 nm	n/a	Photocatalytic oxidation of organics	Higher than P25	Novel route for anatase nanorods	102
Pt/TiNT	Photodeposition	n/a	1 wt%	$CH_3CH_2OH + hv \rightarrow CH_3CHO + H_2$	n/a	Early data on photo-dehydrogenation	126
CdS/TiNT	Ion-exchange surface reaction	6 nm	n/a	Dye oxidation	n/a	Photosensitization of TiNT	117,118
N-doped TiO_2 NP	Calcination with NH_3 (gas)	10 nm	1% (N)	$2 C_3H_6 + 9 O_2 \rightarrow 6 CO_2 + 6 H_2O$	3 times higher than N-P25	Photosensitization of VIS light	113
NiO/TiNF TiNF	Impregnation	n/a	0.2 wt%	$CHCl_3$ oxidation	Higher than P25	Accommodation of Ni in tunnel structure	121

Material	Method	Size		Reaction	Performance	Comments	Ref
TiNT	Microwave	n/a	n/a	$NH_4^+ + O_2 \rightarrow NO_2^- + NO_3^-$	less than P25	Good adsorption of NH_3	99
TiO₂TiNF	Epitaxial growth by precipitation	10–50 nm		Dye oxidation	Higher than TiNF	Support for photocatalyst	108
TiO₂ NF/TiO₂ NP	Acid treatment of TiNF at 180 °C	n/a	n/a	Dye oxidation	Comparable with P25	Novel route for nanostructed anatase	105
Pt/TiNT	sputtering	n/a	n/a	$H_2O + h\nu \rightarrow H_2 + O_2$	Higher than TiO_2	Early data showing water splitting	124
SnTTP/TiNF	*In situ* intercalation	n/a	n/a	Dye oxidation	n/a	Synergy in using UV/VIS light	122
TiNF, TiO₂-(B) NF, anatase NF	Calcination	n/a	n/a	$CH_3OH + h\nu \rightarrow HCHO + H_2$	Higher than P25	Systematic comparison of elongated structures	125
Bi₂Ti₂O₇ NT	AAU template	0.2 μm × ?/μm	n/a	Dye oxidation	Higher than bulk $Bi_2Ti_2O_7$	Effect of dimension on the activity of catalyst	123
TiO₂ NT array	Anodisation	n/a	n/a	$H_2O + h\nu \rightarrow H_2 + O_2$	12–16 %	Perspective electrode with high performance	131
TiO₂ NT array	Anodisation	n/a	n/a	Dye oxidation	Higher than P25	Novel photocatalyst	136

TiNT: titanate nanotubes, TiNF: titanate nanofibres, NR: nanorods, NP: nanoparticles.

a)

b)

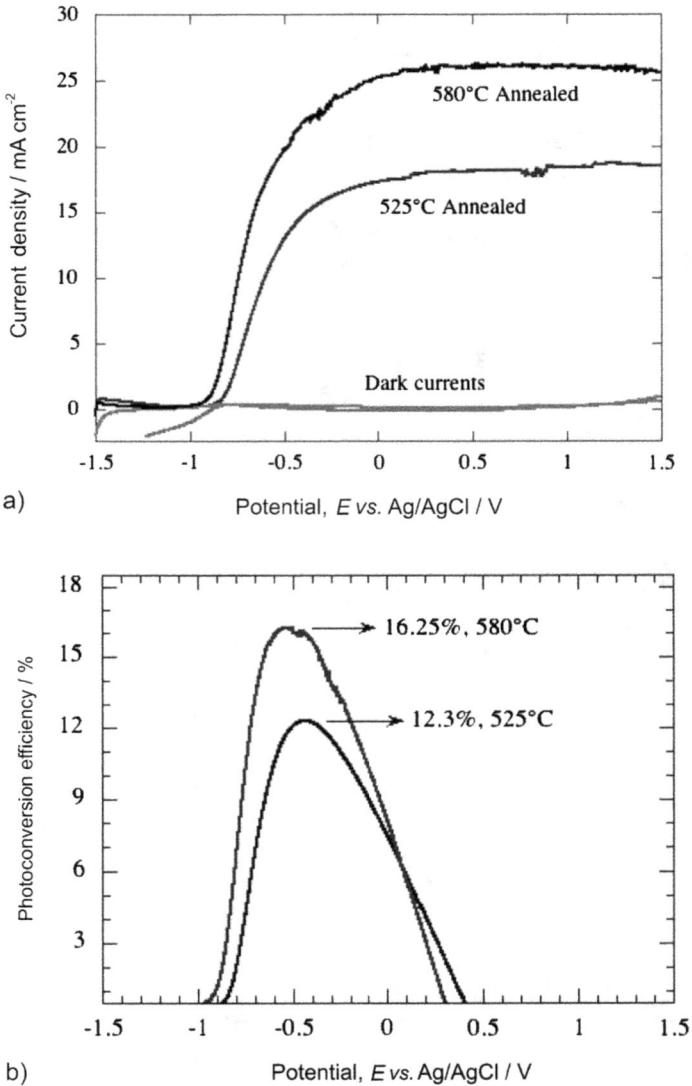

Figure 5.9 Graphs showing: a) photocurrent density and b) the corresponding photo-conversion efficiency of a TiO$_2$ nanotube array electrode with nanotubes of 205 nm outer diameter and 30 μm length. Electrodes were annealed at the indicated temperatures for 1 h in oxygen. Photolysis of water occurred in KOH (1 mol dm^{-3}) electrolyte under UV illumination (98 mW cm^{-2}, 320–400 nm range). (Images are reproduced with kind permission from ref. 131).

evolution under standard conditions), V_{bias} is the applied potential, I_p is the current density responsible for generation of hydrogen and oxygen, and P_t is the power density of illumination ($W\,m^{-2}$).

Figure 5.9 shows that a higher calcination temperature for anodised TiO_2 nanotubes results in an improvement in the photoconversion efficiency and an increase in photocurrent. This is probably due to an improvement in the crystallinity of the nanotube walls leading to a reduction of the amorphous regions and the borders of grain boundaries, which reduces the number of charge-carrier recombination centres. Further increases in calcination temperature, however, result in a deterioration of electrode performance, due to the growth of a barrier layer of oxide which reduces electrical conductivity and a peeling of the nanotubes from the substrate.

Despite relatively good conversion efficiency, TiO_2 nanotube arrays can utilise light of only short wavelength. Another problem with TiO_2 nanotubes is their low catalytic activity in the reaction of oxygen evolution. Both of these problems can be tackled by employing composite materials based on TiO_2 nanotube arrays.

Examples of such composites include Cu–Ti–O (ref. 132) or Fe–Ti–O (ref. 133) ternary oxide nanotube arrays, which can be fabricated by anodic oxidation in a fluoride-containing electrolyte of Cu–Ti and Fe–Ti films, deposited previously by simultaneous co-sputtering. The internal pores of nanotubes can be filled or conformally coated with some narrow band-gap semiconductors leading to heterojuction-type nanotubular composites, including: CdS/TiO_2 NT (ref, 134) and $CdTe/TiO_2$ NT (ref. 135). All of these composites demonstrate a successful sensitisation of composite electrodes to the visible wavelength of incident light, allowing better utilisation of solar radiation.

Anodised TiO_2 nanotube arrays also demonstrate an activity in the photocatalytic oxidation of organic compounds which exceeds the performance of the standard P25 Degussa photocatalyst.[136] This activity can even be further accelerated by decorating the surface of nanotubes with suitable metallic nanoparticles.

5.3 Magnetic Materials

The recent interest in room-temperature ferromagnetic semiconductors and magnetic nanosized materials having a high aspect ratio (motivated by possible applications in spin-base semiconductor devices), has stimulated research into the synthesis and characterisation of elongated titanates and TiO_2 magnetic materials. The pure titanate nanotubes have paramagnetic properties;[137] doping nanotubes with Co^{2+} results in ferromagnetic properties with a coercivity of approximately 40 Oe.[138]

Several methods have been reported for the cobalt doping of titanate nanotubes. The first is *in situ* doping, where cobalt(II) salt is added to a TiO_2–NaOH mixture, followed by alkaline hydrothermal treatment, resulting in the formation of titanate nanotubes.[139] A similar method is the use of Co-doped

TiO$_2$ as a feedstock for the preparation of nanotubes *via* an alkaline hydro-thermal route.[138] In both processes, the crystallization of titanate nanotubes results in the Co(II) ions occupying the octahedral positions of Ti^{4+} in the titanate lattice.[137] In this case, the room temperature ferromagnetism is prob-ably associated with the oxygen vacancies resulting from such substitution.[140]

In contrast, according to previous studies[63] and our own unpublished results, Co^{2+} ion-exchanged titanate nanotubes with a Co : Ti ratio of 2 : 7 possess antiferromagnetic properties at room temperature, probably due to Co–Co interactions. However, it was shown that in the case of Co^{2+} ion-exchanged titanate nanofibres, the hydrothermal treatment of the sample at 160 °C in water, resulted in the substitution of titanium by cobalt and the appearance of ferromagnetic properties.[141] The ferromagnetism decreased at higher Co(II) levels, however, due to the superexchange coupling between cobalt ions in the lattice

The calcination of Co-doped TiNT at 400 and 500 °C results in the formation of Co-doped TiO$_2$-(B) NT and anatase NR, respectively. Both nanostructures are characterised by ferromagnetic properties, with the highest saturation magnetization observed in Co–TiO$_2$-(B) NT.[139] The Co-doped TiNF can also be produced *in situ* by Co(II) addition to a TiO$_2$–NaOH mixture, followed by alkaline hydrothermal treatment at temperatures > 170 °C. The calcination of such Co–TiNF structures at 700 °C results in transformation to anatase nanofibres. Both nanofibre materials have ferromagnetic properties.[39] Quan-tum calculations suggest that Fe(III)-doped titanate nanotubes could possess magnetic insulator properties.[142] Nickel nanoparticles deposited on the surface of TiNTs also show ferromagnetic properties.[143]

5.4 Drug Delivery and Bio-Applications

Recent active areas of research which involve the use of nanostructured inor-ganic materials in biological applications include: controlled drug delivery, the labelling of biological objects, and the building of artificial tissues from nanostructured material composites.[144,145]

Due to their high surface area and affinity towards positively charged ions in aqueous solution, elongated nanostructured titanates have recently been stu-died as a possible element in amperometric bio-sensors. It has been shown that the redox mediator Meldola blue[146] and such oxygen-transport metalloproteins as haemoglobin[147] or myoglobin[148] can be easily immobilized on the surface of TiNT, providing efficient electron transfer between biological molecules and the artificial electrode. Such transfer is usually a challenging task due to the bad compatibility between the inorganic materials and the bio-molecules. The improved charge transfer in such systems can be utilized in bio-sensors, *e.g.* for glucose and NADH.

The studies of ibuprofen adsorption in the pores of TiNT have revealed that the melting temperature of ibuprofen is decreased from 78 °C in the bulk to 66 °C, which is closer to physiological temperatures and might be utilised for controlled drug delivery, following improvements in the technolcgy.[149] The use of titanate nanotubes as capsules for drug delivery and controlled release could be developed based upon a combination of several nanotube properties. A high surface area and large pore volume (due in large extent to the internal cavities of the nanotubes), provide nanotubes with the high load capacity required for drug storage. The nanotubular morphology can be potentially beneficial in drug delivery to the targeted tissue. The brittle nature of the nanotube wall can be utilized for the controlled release of the loaded species stimulated by, for example, ultrasound treatment. Finally, another important property of titanate nanotubes which favours them over, for example, silicon dioxide nanotubes, is their slow dissolution in an acid environment rendering them biodegradable.

Titanium surfaces coated with titanate nanofibres produced by the alkaline hydrothermal method can be used as bio-scaffolds for cell cultures, providing enough rigidity and a large macroporous structure suitable for cell growth and nutrition.[150] The biocompatibility of fibrous sodium titanate deposited inside the pores of anodized TiO_2 nanotube arrays, has been recently demonstrated by observation of the increased growth of hydroxyapatite (HAp) from simulated body fluid.[151] The ability to stimulate crystallisation of HAp on the surface of titanates can be related to their good ion-exchange properties. Hence elongated titanate-coated surfaces are potentially useful in the preparation of well-adhered bioactive surface layers on Ti substrates for orthopaedic and dental implants.

Due to the biocompatibility, the specific range of the pore sizes and the unusual topography of anodized TiO_2 nanotubes, their interactions with biological cells can be important for understanding the *in vivo* processes of titanium-based implants. Recently, the effect of the morphology of TiO_2 nanotube arrays (including diameter and length) on the adhesion, spreading, growth and differentiation of mesenchymal stem cells was thoroughly studied.[152] It was shown that the size of nanotube diameter which most stimulated cell growth and differentiation was approximately 15 nm, while diameters of approximately 100 nm led to drastically increased cell apoptosis.

Further developments in elongated titanates and TiO_2 as materials for biotechnology would require comprehensive studies of their cytotoxicity. Improved methods for the filling of nanotube pores with bioactive drug molecules could provide a significant step towards the development of controlled drug delivery systems.

For many applications, it important to produce nanostructured titanates and titanium dioxides in a form that is suitable for industrial processing. It is also attractive to process materials in a flexible textile form for sensors and transducers. Recent studies have demonstrated the possibility of producing composite, electrically conductive fibres containing titanate nanotubes, carbon nanotubes and chitosan.[153] The material dispersed typically comprised 1 wt% chitosan and 0.3 wt% (SWCN + H-TiNT) mixture in a sonicated aqueous solution. The mixture was coagulated in a bath containing NaOH (10 wt%) in methanol. The presence of SWCNs improved the mechanical, as well as the electrical, properties of the fibres. The resultant fibres showed good bio-compatibility, as evidenced by the adhesion and proliferation of L929 mouse fibroblast cells, and it is suggested that the materials might be suitable for inclusion in biomedical devices such as orthopaedic implants.

5.5 Composites, Surface Finishing and Tribological Coatings

Composite materials based on elongated nanostructures can bring additional functionality and structural reinforcement, allowing the number of applications of such materials to be extended. A few approaches to the manufacture of titanate nanotube–polymer composites have been reported recently using polystyrene[154] or polyurethane[155] matrices. In both cases, the effective dispersion of TiNT was achieved by the hydrophobization of the surface using siloxanes or surface adsorption of hexamethylene diisocyanate, prior to mixing with polystyrene and polyurethane, respectively. TiNT–polystyrene composite films were characterised by an increased Young's modulus and a tensile strength observed even at low nanotube loading levels.[154] Comprehensive tribological studies of TiNT–polyurethane composites have revealed significant improvements in wear resistance and a lower friction coefficient compared to those of the unloaded polymer.[155]

TiNT–polyaniline composites, obtained by the oxidation of aniline adsorbed onto the surface of nanotubes in the presence of triblock copolymers,[156] are interesting due to their unusual combination of electroconductive and proton conductive properties of the polymer and nanotubes, respectively. Similar composites of TiNT with proton conductive Nafion™ ion-exchange polymer have also demonstrated advantages when compared to the pure polymer.[48] Long titanate nanofibres in aqueous suspensions with[157] or without[158] addition of Pluronic F127 fluorocarbon surfactant may form a pulp, which can be cast into a free-standing (paper-like) membrane. These membranes have an open-pore structure (pore size of *ca.* 0.05 μm) and can be used as a filter or catalyst during the oxidation of organic wastes.

The incorporation of nanotubular objects into metal matrices can significantly improve the mechanical properties of composites (see the example of Damascus steel in Chapter 1). One established industrial method for the incorporation of ceramic micro-particles into a metal matrix is co-deposition of particles during electroplating. Our recent results have shown that the embedding of elongated titanate nanotubes into metal during the electroplating of nickel from an electrolyte containing suspended titanate nanotubes, can result in improvement in the wear resistance of the composite coating against steel of approximately 20% (see Figure 5.10).[159]

An important processing requirement for many surface engineering applications is to immobilise titanate nanotubes on the surface of the substrate *via* film formation or the application of a coating. Many approaches have been

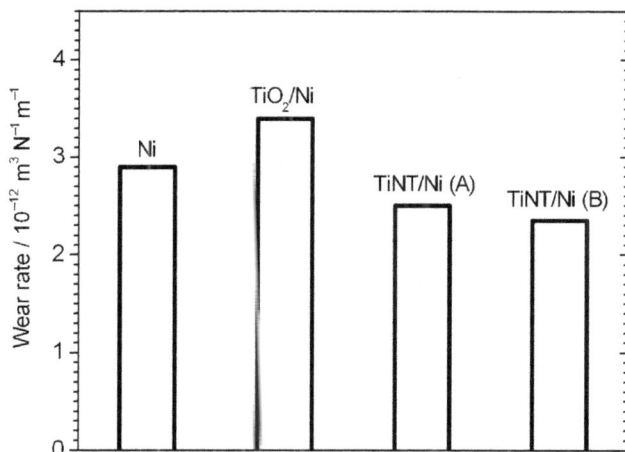

Figure 5.10 Wear resistance of nickel/TiO_2 composites prepared by co-electro-deposition of TiO_2 nanostructures during nickel electroplating. Ni represents electroplated nickel; TiO_2/Ni is TiO_2 (P25) co-deposited from a 20 g dm^{-3} suspension and TiNT/Ni (A) and TiNT/Ni (B) are titanate nanotubes co-deposited from 10 and 20 g dm^{-3} suspensions, respectively. (Data adapted from ref. 159).

used to produce TiNT films, including: the use of the doctor blade technique from a slurry mixture;[6,25] the electrophoretic deposition at the anode from an aqueous electrolyte with the addition of polycations as counter-ions;[127,160] the anodisation of titanium in the presence of fluoride ions;[131] the *in situ* growth on the surface of titanium under alkaline hydrothermal synthesis;[7,8] spin coating;[161] the hot pressing of TiNT powder;[162] Langmuir–Blodgett film deposition;[163] and layer-by-layer assembly.[127,128] All of these methods can produce composite films, characterised by controlled thickness, density and titanate composition, together with a degree of self-assembly.

It is likely that further developments in the synthesis and characterisation of novel composite films will be an active area for future research, particularly in problems related to the self-assembly of elongated nanostructures. This task is challenging, but a favourable outcome would allow significant improvements in the performance of devices such as solar cells.

5.6 Other Applications

A number of recent publications suggest the possible applications of elongated titanates and TiO_2 in specific areas, which are not intrinsic to TiO_2. This includes the use of high surface area elongated titanates as a low-cost adsorbent for chromatography;[164,165] for dye removal from the wastes producing during fabric staining in the textile industry;[3] or for the removal of radioactive ions from waste water.[166]

The high surface area and the acidic nature of the titanate nanotube surface also render these materials useful as a coating for the quartz crystal micro-balance-type devices used for sensing various amines in the gas phase, which can be applied as a detector of chemical warfare agents.[167]

The high aspect ratio of elongated titanates can also be utilized in liquid suspensions, namely nanofluids, which are characterised by unusual electro-rheological[168] and thermoconductive[169] properties. Such fluids can be used in the active control of conventional and intelligent devices where viscosity or thermal conductivity is modulated by an applied electric field.

References

1. D. V. Bavykin and F. C. Walsh, *Eur. J. Inorg. Chem.*, 2009, **8**, 977.
2. D. V. Bavykin, K. E. Redmond, B. P. Wias, A. N. Kulak and F. C. Walsh, *Aust. J. Chem.* 2009, in press.
3. C. K. Lee, C. C. Wang, L. C. Juang, M. D. Lyu, S. H. Hung and S. S. Liu, *Colloids Surf., A*, 2008, **317**, 164.
4. E. Morgado Jr, P. M. Jardim, B. A. Marinkovic, F. C. Rizzo, M. A. S. Abreu, J. L. Zotin and A. S. Araujo, *Nanotechnology*, 2007, **18**, 495710.
5. M. Qamar, C. R. Yoon, H. J. Oh, D. H. Kim, J. H. Jho, K. S. Lee, W. J. Lee, H. G. Lee and S. J. Kim, *Nanotechnology*, 2006, **17**, 5922.
6. S. Uchida, R. Chiba, M. Tomiha, N. Masaki and M. Shirai, *Electrochem.*, 2002, **70**, 418.
7. J. E. Boercker, E. Enache-Pommer and E. S. Aydil, *Nanotechnology*, 2008, **19**, 095604.
8. M. Wei, Y. Konishi, H. Zhou, H. Sugihara and H. Arakawa, *J. Electrochem. Soc.*, 2006, **153**(6), A1232.
9. W. Wang, H. Lin, J. Li and N. Wang, *J. Am. Ceram. Soc.*, 2008, **91**(2), 628.
10. Y. Ohsaki, N. Masaki, T. Kitamura, Y. Wada, T. Okamoto, T. Sekino, K. Niihara and S. Yanagida, *Phys. Chem. Chem. Phys.*, 2005, **7**, 4157.
11. E. Enache-Pommer, J. E. Boercker and E. S. Aydil, *Appl. Phys. Lett.*, 2007, **91**, 123116.
12. P. T. Hsiao, K. P. Wang, C. W. Cheng and H. Teng, *J. Photochem. Photobiol., A*, 2007, **188**, 19.
13. M. S. Akhtar, J. M. Chun and O. B. Yang, *Electrochem. Commun.*, 2007, **9**, 2833.
14. G. K. Mor, O. K. Varghese, M. Paulose, K. Shankar and C. A. Grimes, *Sol. Energy Mater. Sol. Cells*, 2006, **90**, 2011.
15. J. M. Macak, H. Tsuchiya, A. Ghicov, K. Yasuda, R. Hahn, S. Bauer and P. Schmuki, *Curr. Opin. Solid State Mater. Sci.*, 2007, **11**, 3.

16. K. Shankar, J. Bandara, M. Paulose, H. Wietasch, O. K. Varghese, G. K. Mor, T. J. LaTempa, M. Thelakkat and C. A. Grimes, *Nano. Lett.*, 2008, **8**(6), 1654.

17. M. K. Nazeeruddin, A. Kay, I. Rodicio, R. Humphrybaker, E. Muller, P. Liska, N. Vlachopoulos and M. Gratzel, *J. Am. Chem. Soc.*, 1993, **115**, 6382.

18. S. Ngamsinlapasathian, S. Sakulkhaemaruethai, S. Pavasupree, A. Kitiyanan, T. Sreethawong, Y. Suzuki and S. Yoshikawa, *J. Photochem. Photobiol., A*, 2004, **164**, 145.

19. F. Cheng, Z. Tao, J. Liang and J. Chen, *Chem. Mater.*, 2008, **20**, 667.

20. D. V. Bavykin, J. M. Friedrich and F. C. Walsh, *Adv. Mater.*, 2006, **18**, 2807.

21. F. Cheng and J. Chena, *J. Mater. Res.*, 2006, **21**(11), 2744.

22. D. V. Bavykin and F. C. Walsh, *J. Phys. Chem. C*, 2007, **111**, 14644.

23. Y. K. Zhou, L. Cao, F. B. Zhang, B. L. He and H. L. Liz, *J. Electrochem. Soc.*, 2003, **150**, A1246.

24. A. R. Armstrong, G. Armstrong, J. Canales and P. G. Bruce, *J. Power Sources*, 2005, **146**, 501.

25. J. Li, Z. Tang and Z. Zhang, *Electrochem. Commun.*, 2005, **7**, 62.

26. J. Li, Z. Tang and Z. Zhang, *Chem. Phys. Lett.*, 2006, **418**, 506.

27. Y. K. Zhou, L. Cao, F. B. Zhang, B. L. He and H. L. Liz, *J. Electrochem. Soc.*, 2003, **150**, A1246.

28. J. Li, Z. Tang and Z. Zhang, *Chem. Mater.*, 2005, **17**, 5848.

29. X. Gao, H. Zhu, G. Pan, S. Ye, Y. Lan, F. Wu and D. Song, *J. Phys. Chem. B*, 2004, **108**, 2868.

30. L. Kavan, M. Kalbac, M. Zukalova, I. Exnar, V. Lorenzer, R. Nesper and M. Graetzel, *Chem. Mater.*, 2004, **16**, 477.

31. Q. Wang, Z. Wen and J. Li, *Inorg. Chem.*, 2006, **45**, 6944.

32. G. Armstrong, A. R. Armstrong, J. Canales and P. G. Bruce, *Chem. Commun.*, 2005, **19**, 2454.

33. H. Zhang, G. R. Li, L. P. An, T. Y. Yan, X. P. Gao and H. Y. Zhu, *J. Phys. Chem. C*, 2007, **111**, 6143.

34. A. R. Armstrong, G. Armstrong, J. Canales and P. G. Bruce, *Adv. Mater.*, 2005, **17**, 862.

35. M. Zukalova, M. Kalbac, L. Kavan, I. Exnar and M. Graetzel, *Chem. Mater.*, 2005, **17**, 1248.

36. A. R. Armstrong, G. Armstrong, J. Canales and P. G. Bruce, *J. Power Sources*, 2005, **146**, 501.

37. J. Xu, C. Jia, B. Cao and W. F. Zhang, *Electrochim. Acta*, 2007, **52**, 8044.

38. J. Xu, Y. Wang, Z. Li and W. F. Zhang, *J. Power Sources*, 2008, **175**, 903.

39. X. W. Wang, X. P. Gao, G. R. Li, T. Y. Yan and H. Y. Zhu, *J. Phys. Chem., C*, 2008, **112**, 5384.

40. L. P. An, X. P. Gao, G. R. Li, T. Y. Yan, H. Y. Zhu and P. W. Shen, *Electrochim. Acta*, 2008, **53**, 4573.

41. B. L. He, B. Dong and H. L. Li, *Electrochem. Commun.*, 2007, **9**, 425.
42. Z. W. Zhao, Z. P. Guo, D. Wexler, Z. F. Ma, X. Wu and H. K. Liu, *Electrochem. Commun.*, 2007, **9**, 697.
43. J. Kim and J. Cho, *J. Electrochem. Soc.*, 2007, **154**(6), A542.
44. M. Wang, D. J. Guo and H. L. Li, *J. Solid State Chem.*, 2005, **178**, 1996.
45. F. Hu, F. Ding, S. Song and P. K. Shen, *J. Power Sources*, 2006, **163**, 415.
46. H. Song, X. Qiu, D. Guo and F. Li, *J. Power Sources*, 2008, **178**, 97.
47. C. Ponce-de-León, D. V. Bavykin and F. C. Walsh, *Electrochem. Commun.*, 2006, **8**, 1655.
48. B. R. Matos, E. I. Santiago, F. C. Fonseca, M. Linardi, V. Lavayen, R. G. Lacerda, L. O. Ladeira and A. S. Ferlauto, *J. Electrochem. Soc.*, 2007, **154**(12), B1358.
49. L. Schlapbach and A. Züttel, *Nature*, 2001, **414**, 353.
50. A. Zuttel, A. Borgschulte and L. Schlapbach, *Hydrogen as a Future Energy Carrier*, Wiley-VCH, Weinheim, 2008.
51. D. V. Bavykin, A. A. Lapkin, P. K. Plucinski, J. M. Friedrich and F. C. Walsh, *J. Phys. Chem., B*, 2005, **109**, 19422.
52. S. H. Lim, J. Luo, Z. Zhong, W. Ji and J. Lin, *Inorg. Chem.*, 2005, **44**, 4124.
53. K. A. Lokshin, Y. Zhao, D. He, W. L. Mao, H. K. Mao, R. J. Hemley, M. V. Lobanov and M. Greenblatt, *Phys. Rev. Lett.*, 2004, **93**, 125503.
54. X. Wang, J. Zhuang, J. Chen, K. Zhou and Y. Li, *Angew. Chem. Int. Ed.*, 2004, **43**, 2017.
55. O. K. Vardhese, D. Gong, M. Paulose, K. O. Ong, E. C. Dickey and C. A. Grimes, *Adv. Mater.*, 2003, **15**, 624.
56. S. Yoriya, H. E. Prakasam, O. K. Varghese, K. Shankar, M. Paulose, G. K. Mor, T. J. Latempa and C. A. Grimes, *Sens. Lett.*, 2006, **4**, 334.
57. C. A. Grimes, *J. Mater. Chem.*, 2007, **17**, 1451.
58. H. S. Kim, W. T. Moon, Y. K. Jun and S. H. Hong, *Sens. Actuators, B*, 2006, **120**, 63.
59. C. H. Han, D. W. Hong, I. J. Kim, J. Gwak, S. D. Han and K. C. Singh, *Sens. Actuators, B*, 2007, **128**, 320.
60. G. Wang, Q. Wang, W. Lu and J. Li, *J. Phys. Chem. B*, 2006, **110**, 22029.
61. C. H. Lin, S. H. Chien, J. H. Chao, C. Y. Sheu, Y. C. Cheng, Y. J. Huang and C. H. Tsai, *Catal. Lett.*, 2002, **80**(3-4), 153.
62. A. Kleinhammes, G. W. Wagner, H. Kulkarni, Y. Jia, Q. Zhang, L. C. Qin and Y. Wu, *Chem. Phys. Lett.*, 2005, **411**, 81.
63. X. Sun and Y. Li, *Chem. Eur. J.*, 2003, **9**, 2229.
64. D. V. Bavykin, A. A. Lapkin, P. K. Plucinski, L. Torrente-Murciano, J. M. Friedrich and F. C. Walsh, *Top. Catal.*, 2006, **39**(3–4), 151.
65. F. C. Walsh, D. V. Bavykin, L. Torrente-Murciano, A. A. Lapkin and B. A. Cressey, *Trans. Inst. Met. Finish.*, 2006, **84**, 293.
66. B. Zhu, Q. Guo, X. Huang, S. Wang, S. Zhang, S. Wu and W. Huang, *J. Mol. Catal. A: Chem.*, 2006, **249**, 211.

67. B. Zhu, Q. Guo, S. Wang. X. Zheng, S. Zhang, S. Wu and W. Huang, *React. Kinet. Catal. Lett.*, 2006, **88**(2), 301.
68. J. Jiang, Q. Gao and Z. Chen, *J. Mol. Catal. A: Chem.*, 2008, **280**, 233.
69. B. Zhu, K. Li, Y. Feng, S. Zhang, S. Wu and W. Huang, *Catal. Lett.*, 2007, **118**, 55.
70. B. Zhu, X. Zhang, S. Wang. S. Zhang, S. Wu and W. Huang, *Microporous Mesoporous Mater.*, 2007, **102**, 333.
71. S. H. Chien, Y. C. Liou and M. C. Kuo, *Synth. Met.*, 2005, **152**, 333.
72. V. Idakiev, Z. Y. Yuan, T. Tabakova and B. L. Su, *Appl. Catal., A*, 2005, **281**, 149.
73. L. M. Sikhwivhilu, N. J. Coville, D. Naresh, K. V. R. Chary and V. Vishwanathan, *Appl. Catal., A*, 2007, **324**, 52.
74. L. M. Sikhwivhilu, N. J. Coville, B. M. Pulimaddi, J. Venkatreddy and V. Vishwanathan, *Catal. Commun.*, 2007, **8**, 1999.
75. L. Torrente-Murciano, A. A. Lapkin, D. V. Bavykin, F. C. Walsh and K. Wilson, *J. Catal.*, 2007, **245**, 270.
76. J. N. Nian, S. A. Chen, C. C. Tsai and H. Teng, *J. Phys. Chem., B*, 2006, **110**, 25817.
77. D. V. Bavykin, A. A. Lapkin, P. K. Plucinski, J. M. Friedrich and F. C. Walsh, *J. Catal.*, 2005, **235**, 10.
78. M. A. Cortes-Jacome, M. Morales, C. Angeles-Chavez, L. F. Ramirez-Verduzco, E. Lopez-Salinas and J. A. Toledo-Antonio, *Chem. Mater.*, 2007, **19**, 6605.
79. M. A. Cortes-Jacome, C. Angeles-Chavez, M. Morales, E. Lopez-Salinas and J. A. Toledo-Antonio, *J. Solid State Chem.*, 2007, **180**, 2682.
80. C. Hippe, M. Wark, E. Lork and G. Schulz-Ekloff, *Microporous Mesoporous Mater.*, 1999, **31**, 235.
81. Y. Sato, M. Koizumi, T. Miyao and S. Naito, *Catal. Today*, 2006, **111**, 164.
82. J. Qu, X. Zhang, Y. Wang and C. Xie, *Electrochim. Acta*, 2005, **50**, 3576.
83. B. Dong, B. L. He, J. Huang, G. Y. Gao, Z. Yang and H. L. Li, *J. Power Sources*, 2008, **175**, 266.
84. L. Yu and X. Zhang, *Mater. Chem. Phys.*, 2004, **87**, 168.
85. Y. G. Wang, Z. D. Wang and Y. Y. Xia, *Electrochim. Acta*, 2005, **50**, 5641.
86. Y. G. Wang and X. G. Zhang, *Electrochim. Acta*, 2004, **49**, 1957.
87. G. Bo, Z. Xiaogang, Y. Changzhou, L. Juan and Y. Long, *Electrochim. Acta*, 2006, **52**, 1028.
88. F. Tao, Y. Shen, Y. Liang and H. Li, *J Solid State Electrochem.*, 2007, **11**, 853.
89. H. K. Xin, Z. Xiaogang and L. Juan, *Electrochim. Acta*, 2006, **51**, 1289.
90. A. Fujishima, K. Hashimoto and T. Watanabe, *TiO2 Photocatalysis: Fundamentals and Applications*, BKC, USA, 1999.
91. A. Riss, T. Berger, H. Grothe, J. Bernardi, O. Diwald and E. Knolzinger, *Nano Lett.*, 2007, **7**(2), 433.

92. D. V. Bavykin, S. N. Gordeev, A. V. Moskalenko, A. A. Lapkin and F. C. Walsh, *J. Phys. Chem., B*, 2005, **109**, 8565.
93. N. Sakai, Y. Ebina, K. Takada and T. Sasaki, *J. Am. Chem. Soc.*, 2004, **126**, 5851.
94. L. Qian, Z. S. Jin, S. Y. Yang, Z. L. Du and X. R. Xu, *Chem. Mater.*, 2005, **17**, 5334.
95. Q. Li, J. Zhang, Z. Jin, D. Yang, X. Wang, J. Yang and Z. Zhang, *Electrochem. Commun.*, 2006, **8**, 741.
96. T. Tachikawa, S. Tojo, M. Fujitsuka, T. Sekino and T. Majima, *J. Phys. Chem., B*, 2006, **110**(29), 14055.
97. M. Qamar, C. R. Yoon, H. J. Oh, N. H. Lee, K. Park, D. H. Kim, K. S. Lee, W. J. Lee and S. J. Kim, *Catal. Today*, 2008, **131**, 3.
98. C. K. Lee, C. C. Wang, M. D. Lyu, L. C. Juang, S. S. Liu and S. H. Hung, *J. Colloid Interface Sci.*, 2007, **316**, 562.
99. H. H. Ou, C. H. Liao, Y. H. Liou, J. H. Hong and S. L. Lo, *Environ. Sci. Technol.*, 2008, **42**, 4507.
100. G. S. Guo, C. N. He, Z. H. Wang, F. B. Gu and D. M. Han, *Talanta*, 2007, **72**, 1687.
101. M. Zhang, Z. Jin, J. Zhang, X. Guo, J. Yang, W. Li, X. Wang and Z. Zhang, *J. Mol. Catal., A*, 2004, **217**, 203.
102. J. Yu, H. Yu, B. Cheng and C. Trapalis, *J. Mol. Catal., A*, 2006, **249**, 135.
103. Z. Gao, S. Yang, C. Sun and J. Hong, *Sep. Purif. Technol.*, 2007, **58**, 24.
104. J. Yu, H. Yu, B. Cheng, X. Zhao and Q. Zhang, *J. Photochem. Photobiol., A*, 2006, **182**, 121.
105. Y. Yu and D. Xu, *Appl. Catal., B*, 2007, **73**, 166.
106. G. Li, S. Ciston, Z. V. Saponjic, L. Chen, N. M. Dimitrijevic, T. Rajh and K. A. Gray, *J. Catal.*, 2008, **253**, 105.
107. H. L. Kuo, C. Y. Kuo, C. H. Liu, J. H. Chao and C. H. Lina, *Catal. Lett.*, 2007, **113**, 7.
108. H. Yu, J. Yu and B. Cheng, *J. Mol. Catal., A*, 2006, **253**, 99.
109. H. Yu, J. Yu, B. Chenga and J. Lin, *J. Hazard. Mater.*, 2007, **147**, 581.
110. H. Tokudome and M. Miyauchi, *Chem. Lett.*, 2004, **33**(9), 1108.
111. H. Langhuan, S. Zhongxin and L. Yingliang, *J. Ceram. Soc. Jpn.*, 2007, **115**(1), 28.
112. Y. Chen, S. Zhang, Y. Yu, H. Wu, S. Wang, B. Zhu, W. Huang and S. Wu, *J. Dispersion Sci. Technol.*, 2008, **29**, 245.
113. C. Feng, Y. Wang, Z. Jin, J. Zhang, S. Zhang, Z. Wu and Z. Zhang, *New J. Chem.*, 2008, **32**, 1038.
114. H. Yamashita, Y. Ichihashi, M. Takeuchi, S. Kishiguchi and M. Anpo, *J. Synchrotron Radiat.*, 1999, **6**, 451.
115. S. Zhang, Y. Chen, Y. Yu, H. Wu, S. Wang, B. Zhu, W. Huang and S. Wu, *J. Nanopart. Res.*, 2008, **10**, 871.
116. C. C. Tsai and H. Teng, *Appl. Surf. Sci.*, 2008, **254**, 4912.
117. M. Hodos, E. Horvath, H. Haspel, A. Kukovecz, Z. Konya and I. Kiricsi, *Chem. Phys. Lett.*, 2004, **399**, 512.

118. M. W. Xiao, L. S. Wang, Y. D. Wu, X. J. Huang and Z. Dang, *Nanotechnology.*, 2008, **19**, 015706.
119. J. Zhu, D. Yang, J. Geng, D. Chen and Z. Jiang, *J. Nanopart. Res.*, 2008, **10**, 729.
120. H. Li, B. Zhu, Y. Feng, S. Wang, S. Zhang and W. Huang, *J. Solid State Chem.*, 2007, **180**, 2136.
121. H. Song, H. Jiang, T. Liu, X. Liu and G. Meng, *Mater. Res. Bull.*, 2007, **42**, 334.
122. J. H. Jang, K. S. Jeon, S Oh, H. J. Kim, T. Asahi, H. Masuhara and M. Yoon, *Chem. Mater.*, 2007, **19**, 1984.
123. H. Zhou, T. J. Park and S. S. Wong, *J. Mater. Res.*, 2006, **21**(11), 2941.
124. M. Kitano, R. Mitsui, D. Rakhmawaty, E. Zeinhom, M. A. El-Bahy, M. Matsuoka, M. Ueshima and M. Anpo, *Catal. Lett.*, 2007, **119**, 217.
125. J. Jitputti, Y. Suzuki and S. Yoshikawa, *Catal. Commun.*, 2008, **9**, 1265.
126. C. H. Lin, C. H. Lee, J. H Chao, C. Y. Kuo, Y. C. Cheng, W. N. Huang, H. W. Chang, Y. M. Huang and M. K. Shih, *Catal. Lett.*, 2004, **98**(1), 61.
127. M. Miyauchi and H. Tokudome, *Thin Solid Films*, 2006, **515**, 2091.
128. M. Miyauchi and H. Tokudome, *J. Mater. Chem.*, 2007, **17**, 2095.
129. E. Balaur, J. M. Macak, L. Taveira and P. Schmuki, *Electrochem. Commun.*, 2005, **7**, 1066.
130. K. Shankar, J. I. Basham, N. K. Allam, O. K. Varghese, G. K. Mor, X. Feng, M. Paulose, J. A. Seabold, K. S. Choi and C. A. Grimes, *J. Phys. Chem., C*, 2009, **113**, 6327.
131. K. Shankar, G. K Mor, H. E Prakasam, S. Yoriya, M. Paulose, O. K Varghese and C. A Grimes, *Nanotechnology*, 2007, **18**, 065707.
132. G. K. Mor, O. K. Varghese, R. H. T. Wilke, S. Sharma, K. Shankar, T. J. Latempa, K. S. Choi and C. A. Grimes, *Nano Lett.*, 2008, **8**, 1906.
133. G. K. Mor, H. E. Prakasam, O. K. Varghese, K. Shankar and C. A. Grimes, *Nano Lett.*, 2007, **7**, 2356.
134. S. G. Chen, M. Paulose, C. Ruan, G. K. Mor, O. K. Varghese, D. Kouzoudis and C. A. Grimes, *J. Photochem. Photobiol., A*, 2006, **177**, 177.
135. J. A. Seabold, K. Shankar, R. H. T. Wilke, M. Paulose, O. K. Varghese, C. A. Grimes and K. S. Choi, *Chem. Mater.*, 2008, **20**, 5266.
136. J. M. Macak, M. Zlamal, J. Krysa and P. Schmuki, *Small*, 2007, **3**, 300.
137. C. Huang, X. Liu, L. Kong, W. Lan, Q. Su and Y. Wang, *Appl. Phys. A*, 2007, **87**, 781.
138. D. Wu, Y. Chen, J. Liu, X. Zhao, A. Li and N. Ming, *Appl. Phys. Lett.*, 2005, **87**, 112501.
139. X. W. Wang, X. P. Gao, G. R. Li, L. Gao, T. Y. Yan and H. Y. Zhu, *Appl. Phys. Lett.*, 2007, **91**, 143102.
140. S. V. Chong, K. Kadowaki, J. Xia and H. Idriss, *Appl. Phys. Lett.*, 2008, **92**, 232502.
141. H. Zhang, T. Ji, Y. Liu and J. Cai, *J. Phys. Chem., C*, 2008, **112**, 8604.

142. X. G. Xu, X. Ding, Q. Chen and L. M. Peng, *Phys. Rev., B*, 2006, **73**, 165403.
143. J. Jiang, Q. Gao, Z. Chen, J. Hu and C. Wu, *Mater. Lett.*, 2006, **60**, 3803.
144. S. M. Moghimi, A. C. Hunter and J. C. Murray, *Pharmacol. Rev.*, 2001, **53**(2), 283.
145. W. J. Parak, D. Gerion, T. Pellegrino, D. Zanchet, C. Micheel, S. C. Williams, R. Boudreau, M. A. Le Gros, C. A Larabell and A. P. Alivisatos, *Nanotechnology*, 2003, **14**, R15.
146. D. V. Bavykin, E. V. Milsom, F. Marken, D. H. Kim, D. H. Marsh, D. J. Riley, F. C. Walsh, K. H. El-Abiary and A. A. Lapkin, *Electrochem. Commun.*, 2005, **7**, 1050.
147. W. Zheng, Y. F. Zheng, K. W. Jin and N. Wang, *Talanta*, 2008, **74**, 1414.
148. A. Liu, M. Wei, I. Honma and H. Zhou, *Anal. Chem.*, 2005, **77**, 8068.
149. X. P. Tang, N. C. Ng, H. Nguyen, G. Mogilevsky and Y. Wu, *Chem. Phys. Lett.*, 2008, **452**, 289.
150. W. Dong, T. Zhang, J. Epstein, L. Cooney, H. Wang, Y. Li, Y. B. Jiang, A. Cogbill, V. Varadan and Z. R. Tian, *Chem. Mater.*, 2007, **19**, 4454.
151. S. H. Oh, R. R. Finones, C. Daraio, L. H. Chen and S. Jin, *Biomaterials*, 2005, **26**, 4938.
152. J. Park, S. Bauer, K. Von der Mark and P. Schmuki, *Nano Lett.*, 2007, **7**, 1686.
153. C. Dechakiatkrai, C. Lynam, K. J. Gilmore, J. Chen, S. Phanichphant, D. V. Bavykin, F. C. Walsh and G. G. Wallace, *Adv. Eng. Mater.*, 2009, **11**, B55.
154. M. T. Byrne, J. E. McCarthy, M. Bent, R. Blake, Y. K. Gun'ko, E. Horvath, Z. Konya, A. Kukovecz, I. Kiricsi and J. N. Coleman, *J. Mater. Chem.*, 2007, **17**, 2351.
155. H. J. Song, Z. Z. Zhang and X. H. Men, *Eur. Polym. J.*, 2008, **44**, 1012.
156. Q. Cheng, V. Pavlinek, Y. He, C. Li, A. Lengalova and P. Saha, *Eur. Polym. J.*, 2007, **43**, 3780.
157. X. Zhang, A. J. Du, P. Lee, D. D. Sun and J. O. Leckie, *J. Membr. Sci.*, 2008, **313**, 44.
158. W. Dong, A. Cogbill, T. Zhang, S. Ghosh and Z. R. Tian, *J. Phys. Chem., B*, 2006, **110**(34), 16819.
159. C. T. J. Low, D. V. Bavykin, J. O. Bello, S. C. Wang, J. A. Wharton, R. J. K. Wood, K. R. Stokes and F. C. Walsh, in preparation.
160. J. Yu and M. Zhou, *Nanotechnology*, 2008, **19**, 045606.
161. M. Miyauchi and H. Tokudome, *Appl. Phys. Lett.*, 2007, **91**, 043111.
162. T. Kubo, A. Nakahira and Y. Yamasaki, *J. Mater. Res.*, 2007, **22**(5), 1286.
163. M. Takahashi, Y. Okada and K. Kobayashi, *Chem. Lett.*, 2008, **37**(3), 276.
164. H. Niu, Y. Cai, Y. Shi, F. Wei, S. Mou and G. Jiang, *J. Chromatogr., A*, 2007, **1172**, 113.

165. Q. Zhou, Y. Ding, J. Xiao, G. Liu and X. Guo, *J. Chromatogr., A*, 2007, **1147**, 10.
166. D. J. Yang, Z. F. Zheng, H. Yong Zhu, H. W. Liu and X. P. Gao, *Adv. Mater.*, 2008, **20**, 2777.
167. X. Ma, T. Zhu, H. Xu, G. Li, J. Zheng, A. Liu, J. Zhang and H. Du, *Anal. Bioanal. Chem.*, 2008, **390**, 1133.
168. J. Yin and X. Zhao, *Nanotechnology*, 2006, **17**, 192.
169. H. Chen, W. Yang, Y. He, Y. Ding, L. Zhang, C. Tan, A. A. Lapkin and D. V. Bavykin, *Powder Technol.*, 2008, **183**, 63.

Subject Index

www.ingramcontent.com/pod-product-compliance
Lightning Source LLC
Chambersburg PA
CBHW050128240326
41458CB00124B/1703